Advances in Materials Science for Environmental and Energy Technologies

Advances in Materials Science for Environmental and Energy Technologies

Ceramic Transactions, Volume 236

Edited by
Tatsuki Ohji
Mrityunjay Singh
Elizabeth Hoffman
Matthew Seabaugh
Z. Gary Yang

A John Wiley & Sons, Inc., Publication

Published by John Wiley & Sons, Inc., Hoboken, New Jersey.
Published simultaneously in Canada.

For general information on our other products and services or for technical support, please contact our Customer Care Department within the United States at (800) 762-2974, outside the United States at (317) 572-3993 or fax (317) 572-4002.

Wiley also publishes its books in a variety of electronic formats. Some content that appears in print may not be available in electronic formats. For more information about Wiley products, visit our web site at www.wiley.com.

Library of Congress Cataloging-in-Publication Data is available.

ISBN: 978-1-118-27342-5
ISSN: 1042-1122

Printed in the United States of America.

10 9 8 7 6 5 4 3 2 1

Contents

MATERIALS FOR NUCLEAR WASTE DISPOSAL AND ENVIRONMENTAL CLEANUP

ENERGY CONVERSION/FUEL CELLS

Paper presented at the MS&T 2010 meeting in the Materials Solutions for the Nuclear Renaissance symposium.

ENERGY STORAGE: MATERIALS, SYSTEMS AND APPLICATIONS

Preface

The Materials Science and Technology 2011 Conference and Exhibition (MS&T'11) was held October 16–20, 2011, in Columbus, Ohio. A major theme of the conference was Environmental and Energy Issues. Papers from four of the symposia held under that theme are included in this volume. These symposia include Energy Conversion/Fuel Cells; Energy Storage: Materials, Systems and Applications; Green Technologies for Materials Manufacturing and Processing III; and Materials for Nuclear Waste Disposal and Environmental Cleanup. These symposia included a variety of presentations with sessions focused on Fuel Cells & Electrochemistry, Energy Storage, Green Manufacturing and Materials Processing; Waste Minimization; and Immobilization of Nuclear Wastes

The success of these symposia and the publication of the proceedings could not have been possible without the support of The American Ceramic Society and the other organizers of the program. The program organizers for the above symposia is appreciated. Their assistance, along with that of the session chairs, was invaluable in ensuring the creation of this volume.

TATSUKI OHJI, *AIST, JAPAN*
MRITYUNJAY SINGH, *NASA Glenn Research Center, USA*
ELIZABETH HOFFMAN, *Savannah River National Laboratory, USA*
MATTHEW SEABAUGH, *NexTech Materials, USA*
Z. GARY YANG, *Pacific Northwest National Laboratory, USA*

Green Technologies for Materials Manufacturing and Processing

MESOPOROUS MATERIALS FOR SORPTION OF ACTINIDES

Allen W. Apblett
Department of Chemistry, Oklahoma State University,
Stillwater, OK, USA 74078
Zeid Al-Othman
Department of Chemistry, King Saud University
Riyadh 11451, Saudi Arabia

ABSTRACT
The efficient absorption and separation of actinides is of critical importance to numerous aspects of the nuclear industry. For example, uranium extraction from ores and reprocessing of used fuel rods can be significantly simplified and generate less waste by the use of solid actinide extractants. Also, the environmental impact form uranium mining, milling, and extraction activities, the use of spent uranium penetrators, and the legacy of nuclear weapon production can be ameliorated by the use of a highly-efficient adsorbant. Such an adsorbant can also be used to remove uranium from drinking water or ocean water leading to a potentially large increase in uranium reserves. We have developed a mesoporous silica that has significantly enhanced wall-thicknesses and pore sizes that provide improved thermal and hydrothermal stabilities and absorption kinetics and capacities. Grafting of ethylenediamine groups onto the surface using N-[3-(trimethoxysilyl)propyl]ethylenediamine produces extractants that can be used to remove actinides from water.

INTRODUCTION
Uranium is a common contaminant of ground water and can arise from natural and anthropogenic sources. Uranium occurs naturally in the earth's crust and in surface and ground water. When bedrock consisting mainly of uranium-rich granitoids and granites comes in contact with soft, slightly alkaline bicarbonate waters under oxidizing conditions, uranium will solubilize over a wide pH range. These conditions occur widely throughout the world. For example, in Finland exceptionally high uranium concentrations up to 12,000 ppb are found in wells drilled in bedrock [1]. Concentrations of uranium up to 700 ppb have been found in private wells in Canada [2] while a survey in the United States of drinking water from 978 sites found a mean concentration of 2.55 ppb [3]. However, some sites in the United States have serious contamination with uranium. For example, in the Simpsonville-Greenville area of South Carolina, high amounts of uranium (30 to 9900 ppb) were found in 31 drinking water wells [4]. The contamination with uranium is believed to be the result of veins of pegmatite that occur in the area. Besides entering drinking water from naturally occurring deposits, uranium can also contaminate the water supply as the result of human activity, such as uranium mining, mill tailings, and even agriculture [5, 6]. Phosphate fertilizers often contain uranium at an average concentration of 150 ppm and therefore are an important contributor of uranium to groundwater [7]. The Fry Canyon site in Utah is a good example of the dangers of uranium mine tailings. The uranium concentrations measured in groundwater at this site were as high as 16,300 ppb with a median concentration of 840 ppb before remedial actions were taken [8]. Depleted uranium

ammunition used in several military conflicts has also been demonstrated as a source of drinking water contamination [9].

Animal testing and studies of occupationally exposed people have shown that the major health effect of uranium is chemical kidney toxicity, rather than radiation hazards [10]. Both functional and histological damage to the proximal tubulus of the kidney have been demonstrated [11]. Little is known about the effects of long-term environmental uranium exposure in humans but there is an association of uranium exposure with increased urinary glucose, alkaline phosphatase, and ß-microglobulin excretion [12], as well as increased urinary albumin levels [13]. As a result of such studies, the World Health Organization has proposed a guideline value of 2 ppb for uranium in drinking water while the US EPA has specified a limit of 30 ppb.

Current municipal treatment practices are not effective in removing uranium. However, experimentation indicates, that uranium removal can be accomplished by a variety of processes such as modification of pH or chemical treatment (often with alum) or a combination of the two [14]. Several sorbants have been shown to be useful for removal of uranium from water. Activated carbon, iron powder, magnetite, anion exchange resin and cation exchange resin were shown to be capable of adsorbing more than 90% of the uranium and radium from drinking water. However, two common household treatment devices were found not to be totally effective for uranium removal [4].

Besides treatment of well water, there is also a strong need for prevention of the spread of uranium contamination from concentrated sources such as uranium mine tailings. Commonly used aboveground water treatment processes are not cost-effective and do not provide an adequate solution to this problem. However, permeable reactive barriers have been demonstrated to be financially viable and elegant alternatives to active pump and treat remediation systems. Such barriers composed of metallic iron, ferric oxyhydroxide, and bone char phosphate have been designed and proven effective for uranium [8]. Iron metal performed the best and consistently lowered the input uranium concentration by more than 99.9% after the contaminated groundwater had traveled 1.5 ft into the permeable reactive barrier.

In this investigation a functionalized mesoporous silica that had pendant ethylenediamine groups was explored as an adsorbant for the separation and removal of uranium from aqueous solution. Amino-functionalized mesoporous silicas show notable adsorption capacities for heavy metals and transition metals from solution [15]. This investigation also took advantage of a novel mesoporous silica with very large pores, thick walls, and thermal and hydrolytic stability that is superior to conventional mesoporous silicas [16].

EXPERIMENTAL

All reagents were commercial products (ACS Reagent grade or higher) and were used without further purification. Water was purified by reverse osmosis and was deionized before use. OSU-6-W mesoporous silica was synthesized via the procedure previously reported by Al-Othman and Apblett [16].

X-ray powder diffraction (XRD) patterns were recorded on a Bruker AXS D-8 Advance X-ray powder diffractometer using copper K_α radiation. The diffraction patterns were recorded for a 2θ range of 17-70° with a step size of 0.02° and a counting time of 18 seconds per step. Crystalline phases were identified using a search/match program and the PDF-2 database of the International Centre for Diffraction Data [17]. Colorimetry was performed on a Spectronic 200 digital spectrophotometer using 1 cm cylindrical cuvettes. The uranium concentrations in the treated solutions were analyzed at $\lambda = 415$ nm after 5 mL solutions were treated with concentrated nitric acid (1.0 mL) to ensure no speciation of metal ions would interfere with the measurement. For infrared spectroscopic measurements, roughly 10 mg of the sample was mixed with approximately 100 mg FTIR-grade potassium bromide and the blend was finely ground. Spectra in the range of 4000 to 400 cm^{-1} were collected by diffuse reflectance of the ground powder with a Nicolet Magna-IR 750 spectrometer. Typically, 128 scans were recorded and averaged for each sample (4.0 cm^{-1} resolution) and the background was automatically subtracted. Dynamic light scattering (DLS) measurements were performed using aqueous suspensions and a Malvern HPPS 3001 instrument

Preparation of Ethylene Diamine-Derivitized Mesporous Silica

The OSU-6-W mesoporous silica was activated by refluxing 10.0 g of the mesoporous silica in 100 ml of dry toluene for 4 hours under dry atmosphere, followed by washing with 50 ml dry toluene, and drying at 80 °C under vacuum. Next, 6.0 g of the dried material was mixed in 100 ml of dry toluene with twenty milliliters of triethylamine and stirred for around one hour at room temperature and was then filtered off with a fine filter funnel and washed with dry toluene (3 X 50 ml). Subsequently, a mixture of 3.0 g ($\sim$ 50 mmol) of the activated mesoporous silica and a solution of 25 mmol ($\sim$ 5.5 ml) of N-[3-(trimethoxysilyl)propyl]ethylenediamine (TMSPEDA) in 100 ml of dry toluene was heated at relux for 48 hrs under a dry atmosphere. The mixture was then cooled to room temperature and the resulting light brown mixture was filtered with a fine filter funnel. The solid was washed three times with toluene (3 X 50 ml) and then ethanol to rinse away any leftover TMSPEDA. During washing the light brown solid turned white. The white solid was then dried at 80 °C under vacuum for 24 hrs in a Chem Dry apparatus. After drying, the solid product was stirred with 50 ml of distilled water for five hours. The mixture was filtered to recover a white solid that was then washed with dry toluene and dried at 80˚C under vacuum for 24 hours. The silica was reacted at reflux with a fresh solution of 25 mmol ($\sim$5.5 ml) of TMSPEDA in 100 ml of dry toluene using the same procedure as the first treatment. After cooling, the resulting solid was isolated by filtration with a fine filter funnel, washed plentifully with toluene (3 X 50 ml) and ethanol then dried at 80˚C under vacuum for 24 hours. The final product, a white solid, was obtained in a yield of 4.96 g.

Uptake of Uranium and Thorium by the Ethylenediamine-Derivitzed Mesoporous Silica

The adsorption experiments were conducted as follows; five different amounts (25-125 mg) of the functionalized mesoporous silica, were shaken for 4 hours with 10 ml of 100 ppm solutions of Th^{4+} and UO_2^{2+} using 20-ml glass vials for each metal ion separately. Measurement of the metal ion concentration was carried out by allowing the insoluble complex to settle down and filtering an appropriate volume of the supernatant using a 0.45μm syringe filter. Inductively coupled plasma atomic emission spectroscopy (ICP-AES) was then used to measure the concentration of the metals.

RESULTS AND DISCUSSION

The ordered mesoporous silicas, OSU-6-W modified with ethylenediamine functional groups were prepared by a post-synthesis method from the reaction of the ordered mesoporous silica, OSU-6-W, with TMSPEDA according to our previously published procedure [dien paper] Diamine-derivitized mesoporous silicas prepared by this method was obtained in good yield and had excellent surface coverage. The mesoporous silica starting material OSU-6-W, and the sample functionalized with TMSPEDA were characterized by XRD. The resulting diffraction patterns in the range of 1.0-10.0° are shown in Figure 1. The XRD patterns of the derivitized samples show strong (100) peaks and smaller (110) and (200) peak intensities, suggesting that the modification process does not strongly affect the framework integrity of the ordered mesoporous OSU-6-W. It can also be noted that the (100) peak shifted to a higher angle upon derivitization indicating an effective decrease of the pore diameter. This can be understood in terms of the volume excluded as the silylation of the OSU-6-W surface walls takes place. According to the average pore diameter of the material from the surface area measurements (see below), this indicates an average thickening of the walls of about 25.9 Å, an increase that statistically would correspond to at least one extra layer of Si-O-Si homogeneously spread on the original wall (~21.0 Å). It also can be noticed that the d_{100} peak has become broader with the addition of the functional groups, indicating a slight alteration of the ordering of the mesoporous structure.

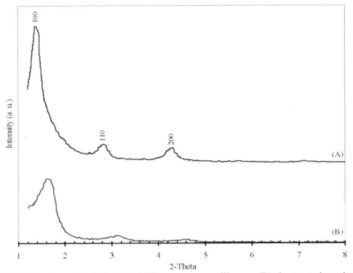

Figure 1. XRD patterns for (A) OSU-6-W mesoporous silica an (B) the treated product. The spectra are shifted vertically for the sake of clarity

Nitrogen adsorption-desorption isotherms were performed at 77 K for the mesoporous silcas. The initial mesoporous silica exhibited a type IV isotherm that showed a sharp, reversible step at ~0.3-0.4 P/P_0, typical of the N_2 filling of uniform mesopores. Analysis of the data gave an average pore diameter of about 51 Å, a pore volume of 1.24 cm³/g and a surface area of 1283 m²/g. After introduction of the functional groups, the nitrogen adsorption-desorption experiments yielded a surface area of 691 m²/g and a total pore volume of 0.58 cm³/g for the derivitized mesoporous silica. The adsorption isotherm curve obtained for the derivitized mesoporous silica showed that the total adsorbed amount of nitrogen (taken at P/P_o=0.99) had diminished, as had the specific surface area. The nitrogen uptake corresponding to the filling of the mesopores was shifted to lower relative pressures indicating a reduction of the pore diameter (from 51.1 to 33.8 Å). An opening of *ca.* 33.8 Å is sufficiently large enough to allow all metal ions and most small organic molecules to be incorporated into the pore channels. The surface area and the total pore volume dropped significantly compared to the un-functionalized sample, OSU-6-W (BET surface area 1283 m^2/g; total pore volume 1.24 cm^3/g). The decrease of the mesoporous volume of the material after silanization is the direct consequence of the silanization process filling the mesopores. However, this quite large decrease probably might also be due to some pore blocking due to partial surface polymerization at the mouths of some mesopores, possibly with silsequioxane polymers. Notably, the pore size distribution (Figure 2) contains three peaks indicating increased heterogeneity of the pore sizes as compared to the more-ordered pores of the mesoporous silica material. Most of the decrease of the specific surface area can be accounted for the uptake of organic species into the pore structure. The textural properties of the mesoporous silica after derivitization are compared to those of the starting materials are provided in Table I.

Table I Textural Properties Determined from Nitrogen Adsorption-Desorption Experiments at 77 K and Powder XRD Measurements.

Sample	Specific surface area (m^2/g)	Total pore volume (cm^3/g)	Average pore size (Å)	d_{100} (Å)	Wall Thickness(Å)
OSU-6-W silica	1283	1.24	51.1	2.4	20.9
Derivitized silica	691	0.58	33.8	1.7	25.9

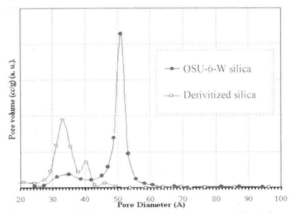

Figure 2. The pore size distribution of OSU-6-W mesoporous silica before and after derivitization.

The presence of covalently linked organic moieties bearing diamine groups in the as-synthesized OSU-6-W mesoporous silicas were also confirmed by ^{13}C CP/MAS solid state NMR spectroscopy [dien paper]. Figure 3 shows the solid state ^{13}C CP/MAS NMR spectra for diamino-functionalized mesoporous silica, The spectrum contained peaks at ^{TM} of 11.4, 23.8, 41.1, 51.6 ppm attributable to the carbon atoms in ($\equiv$Si-CH_2-CH$_2$-), ($\equiv$Si-CH$_2$-CH_2-), ($\equiv$Si-CH$_2$-CH$_2$-CH_2-NH-) and (-NH-CH_2-CH$_2$-NH$_2$), and (-CH_2-NH$_2$), respectively. The lack of peaks attributable to methoxy groups is indicative of complete hydrolysis and condensation of the silane reagent onto the silica surface.

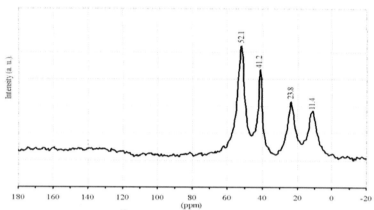

Figure 3 Solid state ^{13}C CP/MAS NMR spectra of the modified samples

The carbon and nitrogen content of the derivitized silica was determined to be 21.56 and 8.59 %, respectively. The concentration of the tethered functional groups was calculated to be 4.63 groups/nm^2 using equation 1 [18].

$$C \text{ (groups/nm}^2) = 6 \times 10^5 \, P_C / [(1200 n_C - W P_C) \, S_{BET}] \tag{1}$$

Where C is the concentration of attached groups, P_C is the percentage of carbon in the sample, n_C is the number of carbon atoms in the attached group (5 for the propylethylenediamine group), W is the corrected formula mass of the modifier ($C_5H_{13}N_2O_3Si$), and S_{BET} is the specific surface area of the pristine substrate (1283 m^2/g).

The adsorption capacity of the modified mesoporous silica toward actinide metal ions was examined using five different amounts of the sorbant ranging from 25 to 125 mg and 100 ppm of UO$_2^{2+}$ ions and Th^{4+} ions. The resulting Langmuir isotherms are shown in Figures 4 and 5. The uptake capacities calculated from the slopes of these plots were 2.22 mmol/g for UO$_2^{2+}$ (528 mg/g) and 2.66 mmol/g for Th^{4+} (616 mg/g). The extremely high capacities correspond to the uptake of 53% by weight of uranyl ions and 62% by weight of thorium ions. Since the density of ethylenediamine groups is 5.32 mmol/g, the uptake can be attributed to the formation of complexes between the metal ions and two of the ethylenediamine ligands.

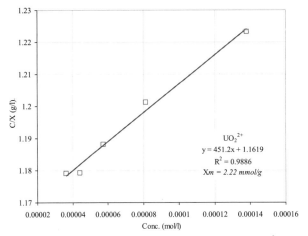

Figure 4. The Langmuir adsorption isotherm for uptake of UO_2^{2+} ions.

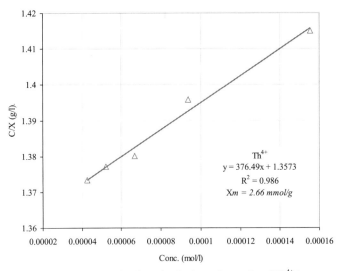

Figure 5. The Langmuir adsorption isotherm for uptake of Th^{4+} ions

The regeneration and reuse of the mesoporous silica was also investigated. Treatment of the uranium-loaded material three times with a stirred aqueous solution of 2.0 M HCl for 1 hour resulted in removal of the bound UO_2^{2+}. Subsequent neutralization of the resulting protonated amines with a dilute ammonium hydroxide returned the derivitized silica to a condition where it could be reused. Regeneration does cause a decrease in the uranium ion uptake capacity (Figure 6). After four adsorption and regeneration cycles the capacity for uranyl ions decreased to approximately 69% of the original value. The decreases may be due to loss of the immobilized groups with washing or a strong interaction of some UO_2^{2+} ions with the amine groups so that it can not be released with HCl washing. Nevertheless, the remaining capacity was still remarkably high.

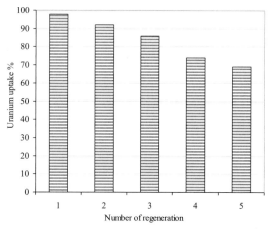

Figure 6. Effect of regeneration of the uptake capacity for UO_2^{2+} ions.

CONCLUSIONS

Mesoporous materials can have a major impact in the area of nuclear and hazardous waste management due to the development of novel materials and processes for rapid adsorption of metals. In this investigation, the reaction of a mesoporous support with a ethylene-diamine containing silylating agent provided a convenient method to prepare a dfunctionalized materaial for sorption of actinide. This sorbant has a well-ordered structure, high surface area, and a high degree of functionalization leading to excellent capacity for removal of uranium and thorium from water.

ACKNOWLEDGEMENT

This paper is based in part upon work supported by the National Science Foundation under Grant Number ECCS-0731208 Any opinions, findings, and conclusions or recommendations expressed in this paper are those of the author(s) and do not necessarily reflect the views of the National Science Foundation. King Saud University is acknowledged for financial support and use of some measurement facilities.

REFERENCES

[1] T. Salonen "238U Series Radionuclides as a Source of Increased Radioactivity in Groundwater Originating from Finnish Bedrock" In: *Future Groundwater Resources at Risk* (International Association of Hydrological Sciences, Wallingford, UK., 1994), 71-84.

[2] M.A. Moss, R.F McCurdy, K.C. Dooley, M.L. Givner, L.C. Dymond, J.M. Slayter, and , M.M. Courneya "Uranium in drinking water - report on clinical studies in Nova Scotia". In: *Chemical Toxicology and Clinical Chemistry of Metals*. S.S. Brown and J. Savory (eds.): Academic Press, London. pp. 149-152 (1983).

[3] United States Environmental Protection Agency "Occurrence and Exposure Assessment for Uranium in Public Drinking Water Supplies" **EPA 68-03-3514** (1990).

[4] J.D. Navratil, "Advances in Treatment Methods for Uranium Contaminated Soil and Water" *Archive of Oncology*, **9**, 257-60 (2001).

[5] C.R. Cothern and W.L. Lappenbusch, "Occurrence of Uranium in Drinking Water in the U.S." *Health Phys.*, **45**, 89 (1983).

[6] D.R. Dreesen, J.M. Williams, M.L. Marple, E.S.Gladney, and D.R. Perrin, "Mobility and Bioavailability of Uranium Mill Tailings Constituents. *Environ. Sci. Technol.*, **16**, 702 (1982).

[7] R.F. Spalding and W.M Sackett "Uranium in Runoff from the Gulf of Mexico Distributive Province: Anomalous Concentrations" *Science*, **175**, 629 (1972)

[8] United States Environmental Protection Agency "Field Demonstration Of Permeable Reactive Barriers To Remove Dissolved Uranium From Groundwater, Fry Canyon, Utah" **EPA 402-C-00-001** (2000)

[9] United Nations Environment Programme "Depleted Uranium in Bosnia and Herzegovina Post-Conflict Environmental Assessment Revised Edition" Switzerland, 2003.

[10] D.M. Taylor and S.K. Taylor "Environmental Uranium and Human Health. *Rev Environ Health* **12**, 147-157 (1997).

[11] D.P. Haley. "Morphologic Changes in Uranyl Nitrate-Induced Acute Renal Failure in Saline- and Water-Drinking Rats. *Lab Investig* **46**,196-208 (1982).

[12] M.L. Zamora, B.L. Tracy, J.M. Zielinski, D.P. Meyerhof and M.A. Moss "Chronic ingestion of Uranium in Drinking Water: a Study of Kidney Bioeffects in Humans. *Toxicological Science*, **43**, 68-77 (1998).

[13] Y. Mao, M. Desmeules, D. Schaubel, D. Bérubé, R. Dyck, D. Brûlé, and B.Thomas, Inorganic Components of Drinking Water and Microalbuminuria. *Environ. Res.*, **71**, 135-140 (1995).

[14] S.K. White, and E.A.Bondietti, Removing Uranium by Current Municipal Water Treatment Processes. *J. Am. Water Works Assoc.*, **75**, 374 (1983)

[15] Z. A. Alothman and A. W. Apblett, "3-Aminopropyltrimethoxysilane functionalized mesoporous materials and uptake of metal ions," *Asian J. Chem.,* **23**[2] 541-546 (2011).

[16] Z. A. Al Othman and A. W. Apblett, "Synthesis and characterization of a hexagonal mesoporous silica with enhanced thermal and hydrothermal stabilities," *Appl. Surf. Sci.,* **256**[11] 3573-3580 (2010).

[17] "Powder Diffraction File (PDF-2)" (International Centre for Diffraction Data, Newtown Square, PA).

[18] L. C. Sander, S. A. Wise, and C. H. Lochmüller, Recent Advances in Bonded Phases for Liquid Chromatography, *Crit. Rev. Anal. Chem.* **18**, 299-417 (1987)

ENVIRONMENTALLY FRIENDLY TIN OXIDE COATING THROUGH AQUEOUS SOLUTION PROCESS

Yoshitake Masuda, Tatsuki Ohji, and Kazumi Kato
National Institute of Advanced Industrial Science and Technology (AIST), Japan
Nagoya 463-8560, Japan

ABSTRACT
This paper gives overviews of two recent achievements of tin oxide nano sheet fabrication through the aqueous solution approaches. One is tin oxide nano-crystals with high surface area, which were synthesized in aqueous solutions at 50°C. The specific surface area reached 194 m^2/g, which was much higher than those of nano-particles in previous studies. A two-dimensional sheet structure was one of the ideal structures for high surface area per unit weight. Novel process allowed us to avoid sintering and deformation of the crystals, and hence realized a high surface area and unique morphology. The other is site-selective tin oxide deposition from aqueous solution, which was performed on super-hydrophilic surfaces. Two-dimensional micro-patterns of tin oxide nano-sheets were fabricated on poly(ethylene terephthalate) (PET) films coated with indium tin oxide (ITO). Tin oxide sheets grew to 100–300 nm in-plane size and 5–10 nm thickness. The site-selective chemical reaction can be applied for precise control of chemical reactions, surface coating, and two-dimensional patterning of tin oxide nanostructures.

INTRODUCTION
Tin oxide has been studied for various applications such as optical device,[1] lithium batteries,[2-5] white pigments for conducting coatings, transparent conducting coatings for furnaces and electrodes, surge arrestors (varistors),[6,7] catalysts,[8,9] opto-conducting coatings for solar cells,[10] etc. Especially, a high surface area of tin oxide is an essential factor to improve device properties of gas sensors, dye-sensitized solar cells, molecular sensors, etc.

Tin oxide particles were prepared by several methods such as precipitation,[5,11] hydrothermal synthesis,[12,13] sol-gel,[14,15] hydrolytic,[16] carbothermal reduction,[17] and polymeric precursor[18] methods. However, high-temperature annealing is required for these processes. It causes an aggregation of the particles, deformation of structures and decrease of surface area. It also increases production cost, energy consumption, CO$_2$ emission, and environmental load. Crystallization in aqueous solutions is one of the smart solutions for these problems. We have demonstrated that tin oxide nano-sheets can be synthesized from proper aqueous solutions and their crystallization was successfully realized below 100°C.[19-22] This paper gives overviews of two recent topics of tin oxide nano sheet fabrication through the aqueous solution approaches; one is tin oxide nano-crystals with a high surface area and unique morphology,[21] and the other is site-selective chemical reaction on flexible polymer films for tin oxide nano-sheet patterning.[22] Crystal growth behavior of tin oxide, SnO and SnO$_2$ in these two cases is also discussed.

HIGH SURFACE AREA OF TIN OXIDE NANO-SHEETS
In order to enhance the reaction, high surface area is required for many applications of tin oxide such as gas sensor, solar cells, and molecular sensors. We have synthesized tin oxide

nano-crystals with high surface area in aqueous solutions at 50°C, and have succeeded in increasing Brunauer-Emmett-Teller (BET) surface area to as high as 194 m^2/g.

SnF_2 (Wako Pure Chemical Industries Ltd., Osaka, Japan, purity 90.0%) (870.6 mg) was dissolved in distilled water (200 mL) at 50°C to be 5 mM. The solutions were kept at 50°C for 20 min and then at 28°C for 3 days without stirring. The nano-crystals precipitated to cover the bottom of the vessels. For comparison, the solutions were centrifuged at 4000 rpm for 10 min after keeping at 50°C for 20 min. Precipitated particles were dried at 60°C for 12 h after the removal of supernatant solutions. SnO_2 and SnO are formed in the aqueous solutions as follows.[19]

$$SnF_2 \overset{OH^-}{\rightarrow} SnF(OH) + F^- \overset{OH^-}{\rightarrow} Sn(OH)_2 + 2F^- \qquad (1)$$
$$Sn(OH)_2 \rightarrow SnO + H_2O \qquad (2)$$
$$Sn(OH)_2 \rightarrow SnO_2 + H_2 \qquad (3)$$
$$SnO + H_2O \rightarrow SnO_2 + H_2 \qquad (4)^{[23,24]}$$

The nano-sheets synthesized at 50°C for 20 min and at 28°C for 3 days were a mixture of SnO_2 as main phase and SnO as additional phase (Fig. 1(a)). For comparison, diffraction pattern of the nano-sheets synthesized at 50°C for 20 min was shown in Fig. 1(b). Full-width half-maximum of the peaks was smaller than that in Fig. 1(a). Lower SnO content resulted in sharp peaks of SnO_2 in Fig. 1(b). These results indicated that SnO mainly formed during the later part of the process (3 days at 28°C), rather than the former part (20 min at 50°C), while SnO_2 was produced from the beginning.

The nano-sheets in supernatant solutions were skimmed with Cu grids for TEM (transmission electron microscopic) observations. They were 20–50 nm in diameter having a uniform thickness (Fig. 2(a)). We observed that they curled up during long-term observation due to electron beams damage, indicating that they have a sheet structure. Some of them tightly connected each other (Fig. 2(b)). They had clear interfaces without pores or small grains. The nano-sheets showed electron diffractions (Fig. 2(b), Inset). Lattice spacing calculated from spots indicated with a red line and a yellow line were 0.283 and 0.154 nm, respectively. The former can be assigned to the 101 crystal plane of SnO_2 (0.264 nm) or 110 crystal plane of SnO (0.269 nm), and the latter can be done to the 310 crystal plane of SnO_2 (0.149 nm), 221 crystal plane of SnO_2 (0.148 nm), 202 crystal plane of SnO (0.149 nm), or 103 crystal plane of SnO (0.148 nm). Additionally, diffraction spots related to the lattice spacing of 0.283 nm were observed at the upper part. They were shown with two white circles and a red line in Fig. 2(b) insert. The double spots indicated that the area shown in a white circle in Fig. 2(b) consisted of two crystals, *i.e.*, two stacked nano-sheets. A high-magnification image was also obtained from the other observation area (Fig. 2(c)). The structure was thinner than that in Figs. 2(a) and (b). A clear image was not obtained in low-magnification images due to low contrast. However, it showed clear high-magnification image and lattice fringe (Fig. 2(c)). Electron diffraction patterns showed a lattice spacing of 0.277 nm (red line), 0.279 nm (green line), and 0.196 nm (yellow line). They were assigned to 110, 1-10, and 200 crystal planes of SnO, respectively. Chemical composition was estimated from several points of tightly packed area. The area included nano-sheets and spherical crystals. The chemical composition varied in the range of Sn:O = 1:1.7–2.7, which was similar to that of SnO_2 rather than SnO. These observations indicated that the crystals were a mixture of SnO_2 and SnO.

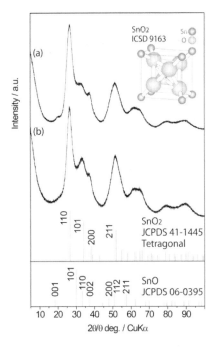

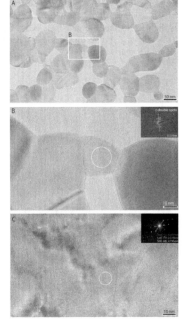

Fig. 1. X-ray diffraction patterns of (a) tin oxide nano-sheets fabricated at 50°C for 20 min and at 28°C for 3 days, and (b) tin oxide nano-sheets fabricated at 50°C for 20 min.[21]

Fig. 2. TEM micrographs and electron diffraction patterns of tin oxide nano-sheets. (b) High-magnification image of (a). (c) High-magnification image of another area.[21]

BET surface area of the nano-sheets was estimated from the N_2 adsorption isotherm (Fig. 3(a)). The nano-sheets had a high surface area of 194 m^2/g (Fig. 3(b)). It was much higher than that of SnO_2 nano-particles such as SnO_2 (BET 47.2 m^2/g, 18.3 nm in diameter, No. 549657-25G, Aldrich), SnO_2 (BET 25.9 m^2/g, 34 nm in diameter, Yamanaka & Co. Ltd., Osaka, Japan), SnO_2 (BET 23 m^2/g, 26 nm in diameter, No. 37314-13, NanoTek, C. I. Kasei Co. Ltd., Tokyo, Japan), and In_2O_3-SnO_2 (BET 3–6 m^2/g, 100–300 nm in diameter, Sumitomo Chemical Co. Ltd.). Furthermore, it was more than double higher than that of the previous report (85 m^2/g).[19] For comparison, BET surface area of the nano-sheets synthesized at 50°C for 20 min was estimated to 146 m^2/g. Pore size distribution was analyzed with Barrett-Joyner-Halenda (BJH) method using adsorption isotherm (Fig. 3(c)). It indicated that the nano-sheet structure included pores of 1–2 nm. Micropore analysis was performed with DFT/Monte-Carlo Fitting (N_2 at 77 K on silica, adsorbent: oxygen), which was completely consistent with isotherms (Fig. 3(d)). Pores of 1–3 nm were shown to be in the nano-sheet structure (Fig. 3(e)). Since such pores were not identified in TEM observation, it can be assumed that these pores are not contained in the nano-sheet crystals themselves, but are gaps between the nano-sheets, which contributed to a high surface area of 194 m^2/g. A two-dimensional sheet structure is one of the ideal structures for high surface area per unit weight.

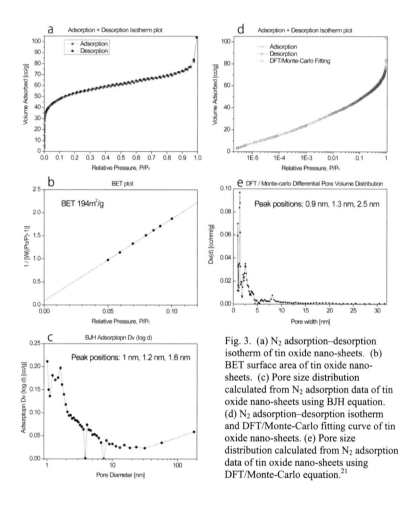

Fig. 3. (a) N_2 adsorption–desorption isotherm of tin oxide nano-sheets. (b) BET surface area of tin oxide nano-sheets. (c) Pore size distribution calculated from N_2 adsorption data of tin oxide nano-sheets using BJH equation. (d) N_2 adsorption–desorption isotherm and DFT/Monte-Carlo fitting curve of tin oxide nano-sheets. (e) Pore size distribution calculated from N_2 adsorption data of tin oxide nano-sheets using DFT/Monte-Carlo equation.[21]

MICRO-PATTERNING OF TIN OXIDE NANO-SHEETS

We have developed a process to realize micro-patterning of tin oxide nano-sheets on flexible polymer films. In this study, flexible polymer films were exposed to light through a photomask to give super-hydrophilic/hydrophobic patterns. In aqueous solutions, tin oxide nano-sheets were crystallized on the super-hydrophilic areas in a site-selective manner to achieve 2D patterning of tin oxide nano-sheets on flexible polymer films.

Poly(ethylene terephthalate) (PET) films were coated with transparent conductive indium tin oxide (ITO) layers. ITO layers of 120–160 nm thickness were coated on the films with the sputtering method. They had a sheet resistance of 10 Ω/square and a transparency of over 74%. They were exposed to vacuum ultraviolet light through a photomask for 10 min (VUV, low-pressure mercury lamp PL16–110, air flow, 100 V, 200 W, SEN Lights Co., 14 mW/ cm^2 for 184.9 nm at a distance of 10 mm from the lamp, 18 mW/ cm^2 for 253.7 nm at a distance of 10 mm the from lamp). SnF$_2$ (Wako Pure Chemical Industries, Ltd., No. 202–05485, FW: 156.71, purity 90.0%) was used as received. Distilled water in polypropylene vessels (200 mL) were capped with polymer films and kept at 90°C. SnF$_2$ (870.6 mg) was added and dissolved in the distilled water at 90°C to a concentration of 5 mM. The substrates were immersed in the middle of the solutions with the bottom up at an angle of 15 degrees from perpendicular. The solutions were stored at 90°C in a drying oven (Yamato Scientific Co., Ltd., DKN402) for 2 h without stirring. The substrates were rinsed under running water and dried with a strong air spray.

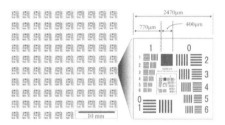

Fig. 4. Design of a photomask for light irradiation.[22]

Fig. 5. (a) Photograph of 2D patterns of tin oxide nano-sheets on a flexible ITO/PET film. (b) Magnified area of (a) showing details of 2D patterns.[22]

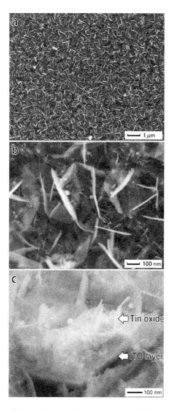

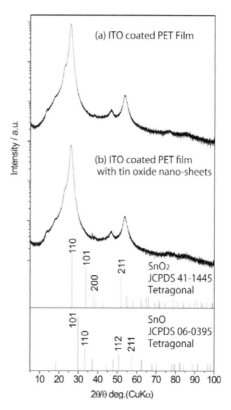

Fig. 6. (a) FE-SEM image of tin oxide nano-sheets on superhydrophilic areas of the film. (b) Magnified area of (a). (c) Tilted image of fracture cross-section of (b).[22]

Fig. 7. XRD patterns of (a) ITO-coated PET film, (b) ITO coated PET film with tin oxide nano-sheets, and JCPDS standard X-ray diffraction data for SnO₂ or SnO.[22]

The flexible PET films coated with indium tin oxide (ITO) initially had a hydrophobic surface with a water contact angle of 111 degrees. The design of photomask shown in Fig. 4 dictated the size and shape of the pattern of the VUV-irradiated surfaces. After irradiation, these surfaces were wetted completely (contact angle 0–5 degrees). In addition, the hydrophilic surface of ITO was much more densely covered with hydroxy groups than the hydrophobic surface of ITO before the irradiation. The initial surfaces of ITO and pure In₂O₃ crystals were covered with adsorbed organic molecules, which increased the water contact angle. The surface hydroxy groups formed chemical bonds with the hydroxy groups on surface of tin oxide nano-crystals, clusters and/or related ions. This bond formation accelerated the nucleation and growth of tin oxide on the surface (Reactions (1) – (2)). The hydrophilic surface was, therefore, more effective for tin oxide deposition than the hydrophobic surface. Generation of gas was observed immediately after the addition of SnF_2 in water at 90°C.

The ion concentration, the valence of the ions, and the pH changed during the synthesis. They affected the formation of SnO and SnO_2. The nano-crystals adhered well to the irradiated surfaces and were not removed under running water or by an air spray. Direct crystallization of tin oxide on the films caused a high adhesive strength, which is required for sensors or solar cells.

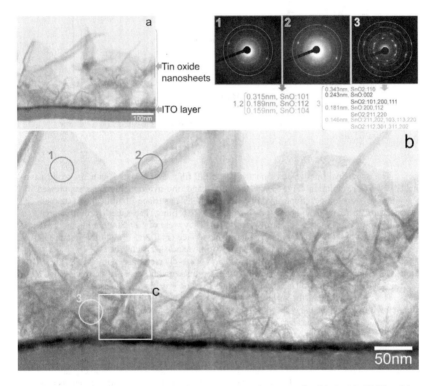

Fig. 8. (a) Cross-section TEM images of tin oxide nano-sheets on a flexible ITO/PET film. (b) Magnified area of (a) showing morphology of tin oxide nano-sheets. (1–3) Electron diffraction patterns from circles in TEM image of (b). [22]

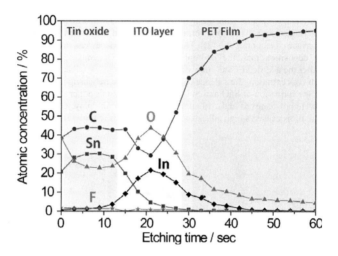

Fig. 9. Atomic concentration depth profile of tin oxide nano-sheets on a flexible ITO/PET film. [22]

The super-hydrophilic areas on the ITO/PET films were coated with tin oxide nano-sheets in aqueous solution (Fig. 5). On the other hand, the hydrophobic areas suppressed the deposition of tin oxide. Super-hydrophilic/hydrophobic surface modification enabled 2D patterning of tin oxide nano-sheets on flexible ITO/PET films. The superhydrophilic area was covered with the nano-sheets uniformly (Fig. 6(a)). They were of 100–300 nm in-plane size and 5–10 nm thickness (Fig. 6(b)). Observation of the surface (Fig. 6(a, b)) and the fractured cross section image (Fig. 6(c)) reveals the morphology of the nano-sheets clearly.

ITO/PET films with tin oxide nano-sheets were characterized with XRD (Fig. 7(b)). Their XRD pattern was similar to that of ITO/PET films without deposited tin oxide (Fig. 7(a)). The tin oxide surface coating was very thin and did not give rise to intense X-ray diffraction peaks. The crystal phase of the surface coatings was evaluated with electron diffraction patterns obtained with TEM. The nano-sheets were formed on ITO layers of PET films directly (Fig. 8), as super-hydrophilic ITO surfaces accelerated the crystal growth of tin oxide. The surface of the ITO layer was covered with nano-sheets of 5–10 nm size (Fig. 8(b)). Some of them further grew to in-plane sizes of 100–300 nm (Fig. 8(b)). Anisotropic growth of tin oxide crystals formed the sheet structure. Lattice fringes were observed from all areas of the sheets (Fig. 8). Electron diffraction patterns revealed that the nano-sheets in circles 1 and 2 were single crystals of SnO. Diffraction spots were clearly observed and assigned to SnO. Estimated lattice spacings matched those of SnO. On the other hand, nano-sheets in circle 3 (Fig. 8) that is close to the ITO surface consisted of SnO and SnO_2, suggesting some amount of SnO_2 crystallized at the initial stage of the immersion period. Similarly to the results of the tin oxide nano-sheets with high surface area, SnO_2 tends to form at the beginning of immersion, while SnO becomes predominant as reaction proceeds. In this study, the reaction is presumably more enhanced due to higher temperature (90°C), resulting in predominance of SnO.

The depth profile from XPS studies reveals the chemical composition of the nano-sheets (Fig. 9). Although the chemical composition of the surface layer high oxygen content, it most likely includes oxygen from surface contamination and lacks sufficient reliability. The ratio of Sn/O changed in 1:0.75–1:0.93 during 3-12 sec etching, corresponding to SnO. This agrees with the analysis results of electron diffraction patterns that the primary crystal phase is SnO.

CONCLUSIONS

This paper gives overviews of two recent achievements of tin oxide nano sheet fabrication through the aqueous solution approaches; one is tin oxide nano-crystals with a high surface area and unique morphology, and the other is site-selective chemical reaction on flexible polymer films for tin oxide nano-sheet patterning.

As for the tin oxide nano-crystals with high surface area, nano-sheets of tin oxides were fabricated in aqueous solutions at ordinary temperature. BET surface area successfully reached 194 m^2/g, which was much higher than that of nano-particles in the previous studies. A two-dimensional sheet structure with pores (or gaps) of 1-3 μm was one of the ideal structures for high surface area per unit weight. Nanosized thickness directly contributed to the high surface area. Crystalline nano-sheets were prepared without high temperature annealing, which degraded the surface area and nanostructures. High surface area and unique nanostructures of the sheets can be applied to gas sensors, dye-sensitized solar cells, and molecular sensors.

Site-selective tin oxide deposition from aqueous solution was performed on super-hydrophilic surfaces. The films were exposed to light through a photomask to form hydrophobic and super-hydrophilic patterns. Chemical reaction was made in a site selective manner only on super-hydrophilic areas in aqueous solutions containing SnF_2. The super-hydrophilic surface of ITO accelerated the chemical reaction, nucleation, and growth of tin oxide crystals, whereas the hydrophobic areas of the films suppressed formation of tin oxide structures. Consequently, two-dimensional patterns of tin oxide nano-sheets were fabricated on PET films coated with ITO. Tin oxide sheets grew to 100–300 nm in-plane size and 5–10 nm thickness. The site-selective chemical reaction can be applied for precise control of chemical reactions, surface coating, and two-dimensional patterning of tin oxide nanostructures.

From results of the two studies, it can be assumed that the obtained tin oxide was a mixture of SnO and SnO_2, and that, as a general tendency, the latter readily formed at the beginning of reaction, while the former became predominant as reaction proceeds. Presumably, in the tin oxide nano-sheets with high surface area, the reaction is sluggish because of low temperature (28°C), and a large amount of SnO_2 remained. On the other hand, in the micro-patterning of tin oxide nano-sheets, the reaction is more enhanced due to higher temperature (90°C), resulting in predominance of SnO. Further study will be performed to elucidate the detailed mechanism of the crystal growth.

ACKNOWLEDGMENTS

This work was partially supported by Ministry of Economy, Trade and Industry (METI), Japan, as part of R&D for High Sensitivity Environment Sensor Components.

REFERENCES

[1]D. S. Ginley and C. Bright, "Transparent Conducting Oxides," *MRS Bull.*, 25 [8] 15–21 (2000).

[2]Y. Idota, T. Kubota, A. Matsufuji, Y. Maekawa, and T. Miyasaka, "Tin-Based Amorphous Oxide: A High-Capacity Lithium-Ion-Storage Material," *Science*, 276 [5317] 1395–7 (1997).

[3]Y. L. Zhang, Y. Liu, and M. L. Liu, "Nanostructured Columnar Tin Oxide Thin Film Electrode for Lithium Ion Batteries," *Chem. Mater.*, 18 [19] 4643–6 (2006).

[4]Y. Wang, J. Y. Lee, and H. C. Zeng, "Polycrystalline SnO_2 Nanotubes Prepared Via Infiltration Casting of Nano-crystallites and their Electrochemical Application," *Chem. Mater.*, 17 [15] 3899–903 (2005).

[5]A. C. Bose, D. Kalpana, P. Thangadurai, and S. Ramasamy, "Synthesis and Characterization of Nano-crystalline SnO_2 and Fabrication of Lithium Cell Using Nano-SnO2," J. Power Sources, 107 [1] 138–41 (2002).

[6]S. A. Pianaro, P. R. Bueno, E. Longo, and J.A Varela, "A New SnO_2-Based Varistor System," *J. Mater. Sci. Lett.*, 14 [10] 692–4 (1995).

[7]P. R. Bueno, M. R. de Cassia-Santos, E. R. Leite, E. Longo, J. Bisquert, G. Garcia-Belmonte, and F. Fabregat-Santiago, "Nature of the Schottky-Type Barrier of Highly Dense SnO_2 Systems Displaying Nonohmic Behavior," *J. Appl. Phys.*, 88 [11] 6545–8 (2000).

[8]T. Tagawa, S. Kataoka, T. Hattori, and Y. Murakami, "Supported SnO2 Catalysts for the Oxidative Dehydrogenation of Ethylbenzene," Appl. Catal., 4 [1] 1–4 (1982).

[9]P. W. Park, H. H. Kung, D. W. Kim, and M. C. Kung, "Characterization of SnO_2/Al_2O_3 Lean NOx Catalysts," *J.Catal.*, 184 [2] 440–54 (1999).

[10]K. L. Chopra, S. Major, and D. K. Pandya, "Transparent Conductors—A Status Review," *Thin Solid Films*, 102 [1] 1–46 (1983).

[11]N. Sergent, P. Gelin, L. Perier-Camby, H. Praliaud, and G. Thomas, "Preparation and Characterisation of High Surface Area Stannic Oxides: Structural, Textural and Semiconducting Properties," *Sens. Actuators B - Chem.*, 84 [2–3] 176–88 (2002).

[12]N. S. Baik, G. Sakai, N. Miura, and N. Yamazoe, *J. Am. Ceram. Soc.*, 83 [12] 2983–7 (2000).

[13]M. Ristic, M. Ivanda, S. Popovic, and S. Music, "Dependence of Nano-crystalline SnO_2 Particle Size on Synthesis Route," *J. Ion-Cryst. Solids* , 303 [2] 270–80 (2002).

[14]L. Broussous, C. V. Santilli, S. H. Pulcinelli, and A. F. Craievich, "SAXS Study of Formation and Growth of Tin Oxide Nano-particles in the Presence of Complexing Ligands," *J. Phys. Chem. B*, 106 [11] 2855–60 (2002).

[15]J. R. Zhang and L. Gao, "Synthesis and Characterization of Nano-crystalline Tin Oxide by Sol– Gel Method," *J. Solid State Chem.*, 177 [4–5] 1425–30 (2004).

[16]Z. X. Deng, C. Wang, and Y. D. Li, "New Hydrolytic Process for Producing Zirconium Dioxide, Tin Dioxide, and Titanium Dioxide Nano-particles," *J. Am. Ceram. Soc.*, 85 [11] 2837– 9 (2002).

[17]E. R. Leite, J. W. Gomes, M. M. Oliveira, E. J. H. Lee, E. Longo, J. A. Varela, C. A. Paskocimas, T. M. Boschi, F. Lanciotti, P. S. Pizani, and P. C. Soares, "Synthesis of SnO_2 Nanoribbons by a Carbothermal Reduction Process," *J. Ianosci. Ianotechnol.* , 2 [2] 125–8 (2002).

[18]E. R. Leite, A. P. Maciel, I. T. Weber, P. N. Lisboa, E. Longo, C. O. Paiva-Santos, A. V. C. Andrade, C. A. Pakoscimas, Y. Maniette, and W. H. Schreiner, "Development of Metal Oxide Nano-particles with High Stability Against Particle Growth Using a Metastable Solid Solution," *Adv. Mater.*, 14 [12] 905–8 (2002).

[19]Y. Masuda and K. Kato, "Aqueous Synthesis of Nano-Sheet Assembled Tin Oxide Particles and their N_2 Adsorption Characteristics," *J. Crys. Growth*, 311, 593–6 (2009).

[20]Y. Masuda and K. Kato, "Tin Oxide Coating on Polytetrafluoroethylene Films in Aqueous Solutions," *Polym. Adv. Technol.*, 21, 211–15 (2010).

[21] Y. Masuda. T. Ohji and K. Kato, "Highly Enhanced Surface Area of Tin Oxide Nano-crystals," *J. Am. Ceram. Soc.*, 93 [8] 2140–43 (2010).

[22] Y. Masuda. T. Ohji and K. Kato, "Site-Selective Chemical Reaction on Flexible Polymer Films for Tin Oxide Nanosheet Patterning," *Eur. J. Inorg. Chem.*, 18, 2819–25 (2011).

[23]C. F. Baes and R. E. Mesiner, *The Hydrolysis of Cations*. John Wiley & Sons Inc., Wiley-Interscience, New York, 1976.

[24]C. Ararat Ibarguena, A. Mosqueraa, R. Parrab, M. S. Castrob, and J. E. Rodriguez-Paeza, "Synthesis of SnO_2 Nano-particles through the Controlled Precipitation Route," *Mater. Chem. Phys.*, 101 [2–3] 433–40 (2007).

INVESTIGATION OF THE MORPHOLOGICAL CHANGE INTO THE FABRICATION OF ZnO MICROTUBES AND MICRORODS BY A SIMPLE LIQUID PROCESS USING Zn LAYERED HYDROXIDE PRECURSOR

Seiji Yamashita, M. Fuji, C. Takai and T. Shirai
Ceramics Research Laboratory, Nagoya Institute of Technology
Honmachi 3-101-1, Tajimi, Gifu 507-0033, Japan

ABSTRACT

Hexagonal-faceted zinc oxide (ZnO) microtubes have been synthesized by a simple liquid process using Zinc Layered Hydroxide precursor. The advantages of this new method are synthesis reaction under a relative low temperature and ambient pressure, and no need for any template or surfactant, as compared to other previous methods. In our procedure, Zinc Layered Hydroxide precursor transformed to ZnO microtube and microrods by heating process in the solution. In this work, in order to study the transformation from precursor to ZnO microtube and microrod, the morphological and crystal structural changes during aging process have been investigated by SEM observation and XRD analysis.

INTRODUCTION

Zinc oxide (ZnO) is a semiconductor with a wide energy band gap (3.37 eV) and a large excitation binding energy (about 60 meV), and has extensive applications for varistor, light emitting diodes, solar cell, catalyst, sensor and cantilever[1]. Recently, synthesis of nano- or micro- ZnO particles with various shapes have been reported such as rod-like[2], flower-like[3,4], tower-like[4], wire-like[5] and tubular[2,5]. These particles have peculiar physical properties derived from their characteristic shapes. For example, well-controlling array of ZnO rod or tower of tube-like particle on a substrate such as silicon, sapphire (α-Al_2O_3) or ITO (indium tin-oxide) is expected to be applied for a photoelectric device and gas sensor element because of their peculiar light-scattering property and high- specific area. The syntheses of these ZnO particles have been demonstrated by vapor phase methods such as thermal evaporation[6], metal organic chemical vapor deposition (MOCVD)[7] and microwave thermal evaporation deposition[8]. Recently, aqueous solution method has been widely investigated for synthesis of ZnO particles with various shapes. The aqueous solution methods have advantages such as simple operation, high cost-efficiency, and hence it would be suitable to large-scale synthesis as compared to the vapor phase methods. The previous work includes use of preseeded substrates[9], surfactant agents such as polyethylene glycol[10] and cetyltrimethylammonium bromide (CTAB)[11] and complexing agent such as citric acid[12].

We have recently developed a new and simple liquid process for synthesis of single-crystalline ZnO microtube, in which ammonia water or bubbles are introduced into $ZnCl_2$ aqueous solution to form white precipitate, followed by aging process, filtration and drying of the precipitate to be ZnO particles[13-15]. The advantages of this new method are synthesis reaction under a relative low temperature and ambient pressure, and no need for any template or surfactant, as compared to other previous methods. In our previous work, the hexagonal layer-like $Zn_5(OH)_{10-y}Cl_y$-H_2O precursor was formed by reaction of $ZnCl_2$ aqueous solution and ammonia on the first step. And the precursor was transformed to ZnO microtubes by heating at high temperature 90 ˚C. According to the previous study, the dissolution of the precursor and re-precipitation to ZnO particles might occur during heating process in the solution and make a change the morphology of ZnO particle[16]. However, the mechanism of the formation of tubular structure has not been understood yet. In this study, we have investigated the effect of the aging time and the drying condition on the morphological change to ZnO microtube and microrod in order to clarify the transformation from precursor to ZnO during heating process in the solution. Furthermore, we discussed the formation mechanism of tubular structure.

EXPERIMENTAL

Aqueous solution of $ZnCl_2$ (0.5 mol/L) was prepared in a conical glass beaker by adding 27.26 g of $ZnCl_2$ (Wako Pure Chemical Industries Ltd.) into 400 mL distilled water and then was heated up to 90 °C under stirrering with a magnetic bar. The pH of $ZnCl_2$ solution was 4.5. Ammonia water (28 %) was added into the $ZnCl_2$ solution until the pH reached 7.5 to form white precipitate. The suspension of precursor aged for a while at 90 °C. After aging process, the precipitate was separated from the solution by a suction filtration with a membrane filter for 30 minutes and then was dried at room temperature 90 °C for 1 day. In order to investigate the effect of drying temperature, the precipitates after filtration and washing by ethanol dried at room temperature (RT) for 1 day. The sample after drying at RT was heated at 90 °C for 1 day. The crystal structure and morphology of the sample of each aging time were characterized by a powder X-ray diffraction (XRD, RINT 1000, Rigaku) and a field emission scanning electron microscope (FE-SEM, JSM-7600F, JEOL).

RESULTS & DISCUSSION

Zinc Layered Hydroxide precursor is hexagonal layered particle with a diameter of 50 μm and a thickness of 1 μm as shown in Figure 1. In our previous work, it has been understood that the crystal structure of the precursor was constituted by the anion exchange of Simonkolleite $[Zn_5(OH)_8Cl_2\text{-}H_2O]$ and it has unstable structure. Therefore, the precursor has a less thermal stability than Simonkolleite and transform to ZnO at 50 °C or more which was lower than that of Simonkolleite [17].

Figure 1 SEM image of Zinc Layered Hydroxide precursor

Figure 2 show SEM images of the obtained particles with different aging time at 90 °C and dried at 90 °C. When the aging time was less than 0.5 hours (Fig 2 (a) and (b)), the tubular particle slightly deposited, but the layered precursor particle still remained. In the sample of aging time for 0.5 and 1 hour, tubular particle was observed. However, when the aging time increased to 2 hour or more (Fig 2 (d), (e) and (f)), the microtube could not be seen and the morphology of obtained particles became rod-like with a diameter of 500 nm and a length of 2~3 μm. In the sample of aging time for 24 hours, large rod-like particles were observed. It is expected that the transformation from precursor to ZnO microtube and microrod occurred mostly less than 1 hour of aging process and tubular particle could be obtained in these aging times.

Figure 2 SEM images of the obtained particles with different aging time of (a) 0h, (b) 0.5 h, (c) 1 h, (d) 2 h, (e) 4 h, (f) 24 h at 90 °C and dried at 90 °C.

In order to characterize the morphology and crystal structure of the sample before heat drying at 90 °C, the samples after aging process were dried at room temperature. Figure 3 shows the SEM images of obtained samples aged at 90 °C and dried at room temperature. Figure 4 shows SEM images of the sample in Fig 3 after heated at 90 °C for 1 day. In Fig 3 (a), it was observed that tubular particles deposited on the layered precursor particles. But the amounts of remaining precursor which could not transform to ZnO were larger than the samples in case of drying at 90 °C (Fig 2 (a)). This means that the morphological and crystal structural change still progressed during drying process at 90 °C. As shown in Figure 5, the XRD patterns of the samples in Fig 3 and 4 exhibit a large remain of precursor in the case of 0.5 hour. When the aging time increased in Fig 3 (b) and (c), peculiar rod-like particles which had double phases of a rough inside phase and smooth external wall were observed. Furthermore, the big growth of the inside phase was seen when the aging time increased, while the growth of the external wall was not observed. In the sample of aging time of 4 hours in Fig 3 (d), most of rod-like particles became dense and peculiar particles were not observed. After heating the samples in Fig 3 at 90 °C, no significant change was observed on the morphology for the samples of aging times of 0.5 and 4 hour (Fig 4 (a) and (d)). However, the pronounced morphological change on the rod-like particles in Fig 4 (b) and (c) can be seen. In Fig 4 (b), the inside phase was decomposed completely while the external wall did not change. Thus, the peculiar rod-like particle became to tubular particle. Thus, it can be thought that the inside phase was easy to decompose by heat treatment because the phase is amorphous. On the other hand, in Fig 4 (c), most of the inside phase remained without decomposition, and therefore, the tubular particles in which the core remained were obtained. It can be considered that the inside amorphous phase of rod-like particle crystallized with the growth by the increase of the aging time. However, the morphology of precursor still remains after heat treatment in both cases. XRD patterns of Fig 5 (a) and (b) revealed that the precursor decomposed with increase of the aging time; it decomposed completely in all the samples after the heat-treatment. In order to observe the inside structure of rod-like particle in Fig 4 (d), the particles were solidified in polymer and immediately sectioned using a LEICA EM U6 ultramicrotome (Leica, U.S.A.). As shown in Fig 6, the cross section of rod-like particle has some defects although the surface of rod-like particle is smooth. Therefore, rod-like particles do not have dense structure inside; they have double phase structure of rough inside phase with defects and smooth external wall.

Figure 3 SEM images of the obtained samples with different aging times of (a) 0.5 h, (b) 1 h, (c) 2 h, (d) 4 h at 90 ˚C and dried at room temperature.

Figure 4 SEM images of prepared samples by heat treatment of the samples in Fig 3 at 90 ˚C.

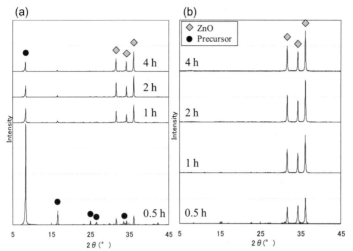

Figure 5 XRD patterns of the samples in (a) Fig 3 and (b) Fig 4.

Figure 6 SEM images of the cutting face of the rod-like particles in Fig 4 (d)

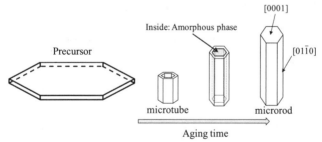

Figure 7 Schematic diagram of transformation from precursor to ZnO microrod.

From the results in this study, transformation mechanism from precursor to ZnO microtube and microrod is considered as shown in Figure 7. In first stage, the dissolution and re-precipitation from precursor is not so fast until 0.5 hour of aging time, therefore, the deposition speed of ZnO particles also slow. During this time, ZnO particles deposit as tubular structure. This tubular formation is most likely since the [0001] polar face is easily dissolved and the [01$\bar{1}$0] non polar face is deposited in this dilute reaction system, because of anisotropic characteristics of ZnO crystal[18]. In the second stage, ZnO particles deposit inside of microtube; however, in this stage (aging time: 1hour or more), the deposited ZnO particles can not crystallize and have many defects because of the fast transformation from the precursor to ZnO. With increase of the aging time, the deposited ZnO particles grow and crystallize, and finally, ZnO microtube become to microrod.

CONCLUSION

It was found that the precursor which has Layered Hydroxide structure, transformed through tubular structure to ZnO microrod during aging process in our study. ZnO microtube was generated in the early stage of the transformation from the precursor. ZnO microtube becomes microrod by deposition inside and crystal growth with increase of the aging time. Observation of the cross section of ZnO microrods revealed that they had a rough inside structure with defects.

REFERENCES
[1] Lukas Schmidt-Mende and Judith L. MacManus-Driscoll, "ZnO – nanostructures, defects, and devices," *Science*, Vol 10, 2007, p 40-48
[2] Quanchang Li, Vageesh Kumar, Yan Li, Haitao Zhang, Tobin J. Marks, and Robert P. H. Chang, "Fabrication of ZnO Nanorods and Nanotubes in Aqueous Solutions" *Chem. Mater.*, Vol 17, 2005, p 1001-1006
[3] Jun Zhang, Lingdong Sun, Jialu Yin, Huilan Su, Chunsheng Liao andChunhua Yan, "Control of ZnO Morphology via a Simple Solution Route" *Chem. Mater.* Vol 14, 2002, p 4172-4177
[4] Zhuo Wang, Xue-feng Qian, Jie Yin and Zi-kang Zhu, "Large-Scale Fabrication of Tower-like, Flower-like, and Tube-like ZnO Arrays by a Simple Chemical Solution Route" *Langmuir*, Vol 20, 2004, p 3441-3448
[5] Jamil Elias, Ramon Tena-Zaera, Guillaume-Yangshu Wang, and Claude Le´vy-Cle´ment, Conversion of ZnO Nanowires into Nanotubes with Tailored Dimensions, *Chem. Mater.* Vol 20, 2008, p 6633–6637
[6] Samuel L. Mensah and Vijaya K. Kayastha, "Formation of single crystalline ZnO nanotubes without catalysts and templates" *Appl. Phys. Lett.* Vol 90, 2007, p 113108-3
[7] B. P. Zhang, N. T. Binh, K. Wakatsuki, and Y. Segawa, "Formation of highly aligned ZnO tubes on sapphire {0001} substrates" *Appl. Phys. Lett.*, Vol 84 [20] 2004, p 4098-4100
[8] Hongbin Chenga, Jiping Chengb, Yunjin Zhangb, Qing-Ming Wang, Large-scale fabrication of ZnO micro-and nano-structures by microwave thermal evaporation deposition, *Journal of Crystal Growth* Vol 299, 2007, p 34–40
[9] Simon P. Garcia, and Steve Semancik, "Controlling the Morphology of Zinc Oxide Nanorods Crystallized from Aqueous Solutions: The Effect of Crystal Growth Modifiers on Aspect Ratio" *Chem. Mater.* Vol 19. 2007
[10] Jinping Liu and Xintang Huang, "A Low-temperature synthesis of ultraviolet-light-emitting ZnO nanotubes and tubular whiskers" *Journal of Solid State Chemistry* Vol 179, 2006, p 843–848
[11] Liming Shen, Ningzhong Bao, Kazumichi Yanagisawa, Kazunari Domen, Craig A. Grimes and Arunava Gupta, "Organic Molecule-Assisted Hydrothermal Self-Assembly of Size-Controled Tubular ZnO Nanostructure" *J. Phys. Chem. C.* Vol 111, 2007, p 7280-7287
[12] Simon P. Garcia and Steve Semancik, "Controlling the Morphology of Zinc Oxide Nanorods Crystallized from Aqueous Solutions: The Effect of Crystal Growth Modifiers on Aspect Ratio" *Chem. Mater.*, Vol. 19, No. 16, 2007

[13] Yong Sheng Han, Li Wei Lin, Masayoshi Fuji, and Minoru Takahashi, "A Novel One-step Solution Approach to Synthesize Tubular ZnO Nanostructures" *Chem. Lett.*, Vol.36, No.8, 2007

[14] Liwei Lin, Yongsheng Han, Masayoshi Fuji, Takeshi Endo, Hideo Watanabe and Minoru Takahashi, "A FACILE METHOD TO SYNT HESIZE ZNO TUBES BY INVOLVING AMMONIA BUBBLES" *Ceramic Transactions*, Vol 198, 2007, p 269-274

[15] Liwei Lin, Yongsheng Han, Masayoshi Fuji, Takeshi Endo, Xiaowei Wang and Minoru Takahashi, "A Novel One-step Solution Approach to Synthesize Tubular ZnO Nanostructures" *J. Ceram. Soc. Jpn.*, Vol 116, 2008, p 168-200

[16] Liwei Lin, Hideo Watanabe, Masayoshi Fuji and Minoru Takahashi, "Morphological Control of ZnO particles synthesized via a New and Facile Aqueous Solution Route, *Advance Powder Technology*, Vol 20, 2009

[17] Seiji Yamashita, Hideo Watanabe, Takashi Shirai, Masayoshi Fuji , Minoru Takahashi, Liquid phase synthesis of ZnO microrods highly oriented on the hexagonal ZnO sheets, *Advanced Powder Technology*, Vol 22, 2011, p 271–276

[18] Wen-Jun Li, Er-Wei Shi, Wei-Zhuo Zhong, Zhi-Wen Yin, Growth mechanism and growth habit of oxide crystals, *Journal of Crystal Growth*, Vol 203, 1999, p 186-196

FABRICATION OF SOLID ELECTROLYTE DENDRITES THROUGH NOVEL SMART PROCESSING

Soshu Kirihara, Satoko Tasaki and Hiroya Abe
Joining and Welding Research Institute, Osaka University
Ibaraki, Osaka, Japan

Katsuya Noritake and Naoki Komori
Graduate School of Engineering, Osaka University
Suita, Osaka, Japan

ABSTRACT

Solid electrolyte dendrites of yttria stabilized zirconia sponges with spatially ordered porous structures were fabricated for fuel cell electrodes through novel smart processing. Micro lattices of 200 µm in diameter composed of acrylic resin with ceramic particles were created by micro patterning stereolithography of computer aided design and manufacturing. Sintered ceramic lattices of 98 % in relative density were obtained through dewaxing and sintering processes. These ceramic lattices with coordination numbers 4, 6 and 8 were free formed. The aspect ratios were modified between 1.0 and 2.0 to value the porosities from 50 to 80 %. Gaseous fluid profiles and pressure distributions in the formed ceramic lattices were visualized and analyzed by using finite difference time domain simulations. The solid electrolyte dendrites with the extremely high porosities and wide surface areas are expect to be applied to novel electrodes in the compact fuel cells.

INTRODUCTION

Novel clean energy generators are developed aggressively to realize the next generation of sustainable society. Solid oxide fuel cells (SOFCs) are investigated as novel generation systems of electric powers with high efficiencies in energy conversion circulations. Yttria stabilized zirconia (YSZ) with high ion conductivities for incident oxygen is widely adopted material for solid electrolyte anodes as the SOFC components[1-5]. These material compositions and microstructures were modified and optimized to obtain the higher performance in the minimized devise system. Recently, random porous structures have been introduced into the YSZ electrodes in micrometer or nanometer sizes to increase surface areas of reaction interfaces and gap volumes of stream paths[6-8]. And, trinary phase points of the ceramic and metal electrolytes toward the reaction gases were formed and increased by using nanoparticles assembling techniques[9,10]. In this investigation, novel solid electrolyte dendrites composed of YSZ spatial lattice structures with various coordination numbers were successfully fabricated by using micro pattering stereolithography and powder sintering techniques. The dendrite structures having the special propagations of the ceramic electrolyte lattices with the ordered geometric pattern were inspired from neuron networks in human brains to transfer electric signals. In the dendrite structures, stress distributions and fluid flows were simulated and visualized by using finite element methods. These new computer aided designs, manufactures and evaluations have been established and optimized to create micro components of various ceramics in our investigation group[11-23].

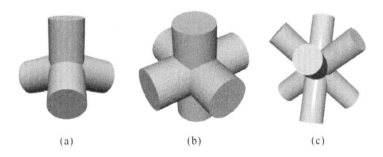

(a)   (b)   (c)

Figure 1: Computer graphics of lattice distributions of dendrite structures (a) (b) and (c) with coordination number 4, 6 and 8, respectively. The volume fractions are changed with aspect ratios.

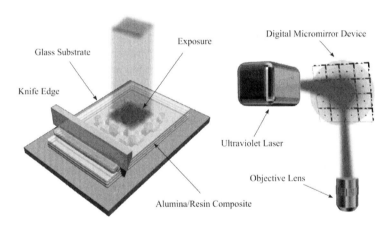

Figure 2: A schematic illustration of micro patterning stereolithography. Fine images are exposed by using a digital micro mirror device on a photo sensitive resin.

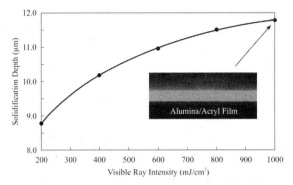

Figure 3: Variation of solidification depth of acryl films with alumina particles according to the visible ray intensities formed by one shot DMD exposure in the micro stereolithography process.

EXPERIMENTAL PROCEDURES

The spatial lattice propagations in the solid electrolyte dendrites were designed by using a computer graphic application (Think Design: Toyota Caelum, Japan) as shown in Figure 1. These surface areas of reaction interfaces and the gap volume of stream paths were calculated geometrically for the dendrite lattice with 4, 6 and 8 coordination numbers. The dendrite lattices of 4 and 1.16 in coordination number and aspect ratio can be considered to exhibit the higher reaction efficiencies and gas transmittances according to Nernst equation. In the optimized dendrite structure, the diameter and length of YSZ rods were decided 92 and 107 μm, respectively. The lattice constant was 250μm. The graphic data was converted into a stereolithographic equipment (D-MEC: SI-C1000, Japan). Photo sensitive acrylic resin dispersed with YSZ particles of 60 and 100 nm in first and second diameters at 30 volume % were fed over a substrate. The thickness of each layer was controlled to 10 μm. The cross sectional pattern was formed through illuminating visible laser of 405 nm in wavelength on the resin surface by applying a digital micro mirror device (DMD). Figure 2 shows a schematic illustration of the micro patterning stereolithography system. The solid micro structures were built by stacking these patterned layers. In order to avoid deformation and cracking during dewaxing, careful investigation for the heat treatment processes were required. The formed precursors were heated at various temperatures from 100 to 600 °C while the heating rate was 1.0 °C /min. The dewaxing process was observed in respect to the weight and color changes. The YSZ particles could be sintered at 1500 °C for 2 hs. The heating rate was 8.0 °C/min. The density of the sintered sample was measured by using Archimedes method. The ceramic microstructures were observed by a digital optical microscope and scanning electron microscopy. In the lattice dendrites, fluid flow velocities and pressure stress distributions were simulated by a finite volume method (FVM) application (Ansys: Cybernet Systems, Japan).

Figure 4: Acryl dendrite lattices including with YSZ particles fabricated by using the micro pattering stereolithography process.

Figure 5: The nanometer sized YSZ particles in the acryl lattices. Particle size and volume fraction are 100 nm and 30 %, respectively.

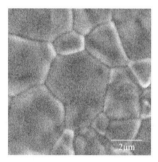

Figure 6: Sintered dendrite lattices of YSZ solid electrolyte. The realized part accuracy of the lattice is 2 μm approximately.

Figure 7: A microstructure of YSZ lattice in the dendrite structure. Average grain size was measured by using SEM at 4 μm.

RESULTS AND DISCUSSION

The acrylic resin films with the alumina particles dispersion were formed by one shot visible lay exposure through the DMD to measure these thicknesses indicating solidification depths through photo polymerizations. The solidification depths were increased continuously according to the visible lay intensities as shown in Figure 3. In the cross sectional microstructures of the formed acrylic films, the alumina particles were dispersed homogeneously. The visible lay intensity was optimized at @ mJ/mm^2 to obtain the solid components through the continuous layer stacking and joining. The dendrite lattice structures composed of the YSZ dispersed acrylic resins were processed exactly by using the stereolithography as shown in Figure 4. The spatial resolution was approximately 1.0 %. The microstructure of the composite lattice is shown in Figure 5. The nanometer sized YSZ particles were dispersed homogeneously in the acryl matrix. Figure 6 shows the sintered solid electrolyte dendrite with the YSZ micro lattice structure. The deformation and cracking were not observed. The linear shrinkages on the both horizontal and vertical axis were 32 %. The volume fraction of the air gaps was 50 % by the open paths. In the other previous investigations, the porous electrodes were formed by sintering the YSZ surly with polystyrene particles dispersion. Therefore, it is difficult to realize the prefect opened pores structures with the higher porosity over 40 % in volume fraction. Figure 7 shows the dense microstructure of the YSZ lattice. The average grain size was approximately 4 μm. The relative density reached 95 %. Micrometer sized cracks or pores were not observed. The obtained dense YSZ lattice structure will exhibit the higher performances in the mechanical properties as the porous electrodes of the solid electrolyte dendrites. The fluid flow velocities were simulated by using the FVM method. Continuously curved lines indicate the fluid distributions along the vector directions of flow velocities. All air paths were opened for outsides and connected with each others in the YSZ dendrite lattice structures. The fluid flows can transmit the one direction smoothly. The pressure stress distributions in the dendrite were simulated. The fluid pressures were gradually distributed for flow direction, and the localization of the stress was not observed. The fabricated solid electrolyte dendrites with YSZ lattice structures can be considered to have higher performances as novel ceramic electrodes in near future SOFCs.

CONCLUSION

Solid electrolyte dendrites with yttria stabilized zirconia lattices were fabricated successfully for anode electrodes of solid oxide fuel cells. Ceramics lattices with various coordination numbers were created and modified to realize larger scale reaction interfaces through computer aided design, manufacture and evaluation techniques. Acryl precursors including ceramic particles were formed successfully by using micro patterning stereolithography. Thorough careful optimization of process parameters in dewaxing and sintering, dense ceramic micro components were obtained. These solid electrolyte dendrites with opened air path networks exhibited effective transmission properties of fluid flows in computer simulations and visualizations. The ceramic electrodes have potentials to contribute for developments of compact fuel cells in the next generation industry.

ACKNOWLEDGMENTS

This study was supported by Priority Assistance for the Formation of Worldwide Renowned Centers of Research - The Global COE Program (Project: Center of Excellence for Advanced Structural and Functional Materials Design) from the Ministry of Education, Culture, Sports, Science and Technology (MEXT), Japan.

REFERENCE

[1] N. Q. Minh, Ceramic Fuel Cells, *J. Am. Ceram.*, **76**, 563-88 (1993).

[2] S. C. Singhal, Solid Oxide Fuel Cells for Stationary, Mobile, and Military Applications, *Solid State Ionics*, **152-153**, 405-10 (2002).

[3] A. Stambouli and E. Traversa, Solid Oxide Fuel Cells: a Review of an Environmentally Clean and Efficient Source of Energy, *Renew. Sust. Energy Rev.*, 6, 433-55 (2002).

[4] T. Ramanarayanan, S. Singhal and E. Wachsman, High Temperature Ion Conducting Ceramics, *Electrochem. Soc. Int.*, **10**, 22-7 (2001).

[5] J. Will, A. Mitterdorfer, C. Kleinlogel, D. Perednis and L. Gauckler, Fabrication of thin electrolytes for second-generation solid oxide fuel cells, *Solid State Ionics*, **131**, 79-96 (2000).

[6] J. Hua, Z. Lüa, K. Chena, X. Huanga, N. Aia, X. Dub, C. Fub, J. Wanga and W. Su, Effect of Composite Pore-former on the Fabrication and Performance of Anode-supported Membranes for SOFCs", *J. Memb. Sci.*, **318**, 445-51 (2008).

[7] J. Haslam, A. Pham, B. Chung, J. Dicarlo and R. Glass, Effects of the Use of Pore Former on Performance of an Anode Supported Solid Oxide Fuel Cell, *J. Am. Ceram.*, **88**, 513-518 (2005).

[8] T. Talebi, M. Sarrafi, M. Haji, B. Raissi and A. Maghsoudipour, Investigation on Microstructures of NiO-YSZ Composite and Ni-YSZ Cermet for SOFCs", *Int. J. Hydrogen Energy*, **35**, 9440-7 (2010).

[9] F. Takehisa, K. Murata, C. Huang, M. Naito, H. Abe and K. Nogi, Morphology Control of SOFC Electrodes by Mechano-Chemical Bonding Technique, *Ceram. Eng. Sci. Proc.*, **25**, 263-267 (2004).

[10] F. Takehisa, K. Murata, S. Ohara, H. Abe, M. Naito and K. Nogi, Morphology Control of Ni-YSZ Cermet Anode for Lower Temperature Operation of SOFC, *J. Power Sources*, **125**, 17-21 (2004).

[11] W. Chen, S. Kirihara and Y. Miyamoto, Fabrication and Measurement of Micro Three-dimensional Photonic Crystals of SiO_2 Ceramic for Terahertz Wave Applications, *J. Am. Ceram.*, **90**, 2078-81 (2007).

[12] W. Chen, S. Kirihara and Y. Miyamoto, Three-dimensional Microphotonic Crystals of ZrO_2 Toughened Al_2O_3 for Terahertz wave applications, *Appl. Phys. Let.*, **91**, 153507-1-3 (2007).

[13] W. Chen, S. Kirihara and Y. Miyamoto, Fabrication of Three-dimensional Micro Photonic Crystals of Resin-Incorporating TiO_2 Particles and their Terahertz Wave Properties, *J. Am. Ceram.*, **90**, 92-6 (2007).

[14] W. Chen, S. Kirihara and Y. Miyamoto, Static Tuning Band Gaps of Three-dimensional Photonic Crystals in Subterahertz Frequencies, *Appl. Phys. Let.*, **92**, 183504-1-3 (2008).

[15] H. Kanaoka, S. Kirihara and Y. Miyamoto, Terahertz Wave Properties of Alumina Microphotonic Crystals with a Diamond Structure, *J. Mat. Res.*, **23**, 1036-41 (2008).

[16] Y. Miyamoto, H. Kanaoka and S. Kirihara, Terahertz Wave Localization at a Three-dimensional Ceramic Fractal Cavity in Photonic Crystals, *J. Appl. Phys.*, **103**, 103106-1-5 (2008).

[17] S. Kirihara, Y. Miyamoto, K. Takenaga, M. Takeda and K. Kajiyama, Fabrication of Electromagnetic Crystals with a Complete Diamond Structure by Stereolithography, *Solid State Comm.*, **121**, 435-9 (2002)

[18] S. Kirihara, M. W. Takeda and K. Sakoda, Y. Miyamoto, Control of Microwave Emission from Electromagnetic Crystals by Lattice Modifications, *Solid State Comm.*, **124**, 135-9 (2002).

[19] S. Kanehira, S. Kirihara and Y. Miyamoto, Fabrication of TiO_2-SiO_2 Photonic Crystals with Diamond Structure, *J. Am. Ceram*, **88**, 1461-4 (2005).

[20] S. Kirihara and Y. Miyamoto, Terahertz Wave Control Using Ceramic Photonic Crystals with Diamond Structure Including Plane Defects Fabricated by Micro-stereolithography, *Int. J. Appl. Ceram. Tech.*, **6**, 41-4 (2009).

[21] S. Kirihara, T. Niki and M. Kaneko, Terahertz Wave Behaviors in Ceramic and Metal Structures Fabricated by Spatial Joining of Micro-stereolithography, *J. Phys.*, **65**, 12082-1-6 (2009).

[22] S. Kirihara, T. Niki and M. Kaneko, Three-dimensional Material Tectonics for Electromagnetic Wave Control by Using Micoro-stereolithography, *Ferroelectrics*, **387**, 102-11 (2009).

[23] S. Kirihara, K. Tsutsumi and Y. Miyamoto, Localization Behavior of Microwaves in Three-dimensional Menger Sponge Fractals Fabricated from Metallodielectric Cu/polyester Media, *Sci. Adv. Mat.*, **1**, 175-81 (2009).

MICROSTRUCTURAL AND MECHANICAL PROPERTIES OF THE EXTRUDED α-β DUPLEX PHASE BRASS Cu-40Zn-Ti ALLOY

H. Atsumi[1], H. Imai[2], S. Li[2], K. Kondoh[2], Y. Kousaka[3] and A. Kojima[3]

[1] Graduate School of Engineering, Osaka University, 565-0871, 2-1, Yamadaoka, Suita, Osaka, Japan
[2] Joining and Welding Research Institute, Osaka University, 567-0047, 11-1, Mihogaoka, Ibaraki, Osaka, Japan
[3] San-etsu Metals Co. Ltd., 939-1315, 1892, Ohta, Tonami, Toyama, Japan

ABSTRACT
    The effect of thermo-mechanical treatment conditions such as solid solutionizing and following hot extrusion on microstructural and mechanical properties of α-β duplex phase Cu-40mass%Zn with a small addition of titanium (Ti) were investigated in the present study. Cu-40Zn-0.5mass%Ti cast ingot was pre-heated at 973 K for 15 min in solid solutionizing and immediately extruded to fabriacate a rod specimen with 7 mm in diameter. The extruded specimen consisted of fine and uniform α-β phases with an average grain size of 2.51 μm. Fine precipitation particles having about 0.5 μm in diameter were also dispersed in both phases. Particularly, α-phase consisted of fine grains due to the pining effect by such fine precipitation particles at the grain boundaries. The tensile property of the extruded sample was YS: 304 MPa, UTS: 543 MPa, and 44 % elongation. The high strengthening mechanism of the wrought brass alloy was mainly due to the grain refinement of α and β phases.

INTRODUCTION

    Copper-Zinc alloys (Cu-Zn: brass) are widely applied in various fields, such as electronic equipment connectors, device components, water pipes of water system, water faucet, valves, pipe connections and hydraulic valves used in automobile, fire control and airplane, because of their excellent electrical and thermal conductivities, corrosion resistance, formability, and machinability. In particular, lead (Pb) is added to the traditional brasses to improve their machinability. However, Pb addition in the materials is significant severe hazard to the environment and human health. The material designs to safety the representative regulation as Restriction of Hazardous Substance (RoHS) and Waste Electrical and Electronic Equipment (WEEE) have been considered.

    Furthermore, since brass alloys having α-β duplex phase structures reveal suitable strength and elongation[1-4], α-β duplex phase brasses (e.g. Cu-40mass%Zn alloy) have been also applied for many industrial products. These alloys were also studied to improve their mechanical properties. For example, the commercial high-strength brass is invented and has been used for marine propellers and bridge bearings because of high mechanical properties and corrosion resistance. This material is alloyed with Al, Mn, and Fe for solid solution strengthening and increasing area fraction of hard β-phase[5-7]. However, when the alloys are added with the larger amounts of elements, the coarser and brittle intermetallic compounds (IMCs) are produced in the matrix, and result in the drastic decrease of machinability. The IMCs drastically decrease their characters such as machinability[8-10]. Thus, brass alloys need to be strengthened by the small amount of additives due to decreasing the content of such IMCs or dispersing the fine IMCs[11-14].

    In this study, high strength α-β duplex phase brass Cu-40Zn with small addition of 0.5 mass% Titanium (Ti) was prepared by casting process, and microstructures and mechanical properties of solid solutionized and immediately extruded alloy were investigated.

EXPERIMENTAL PROCEDURES

Cu-40mass%Zn brass alloy (Cu-40Zn) and Cu-40Zn with addition of 0.5 mass% Ti (Cu-40Zn-0.5Ti) were prepared by casting process. The chemical compositions of brass cast ingots are shown in Table 1. The cast ingots with 60 mm in diameter were machined into billets with 41 mm in diameter. The billets were hot-extruded by the hydraulic press machine (SHP-200-450: Shibayamakikai Co.). Before extrusion, the billet was preheated at 923 K for 3 min and at 973 K for 15 min in solid solution treatment condition in Ar gas atmosphere by muffle furnace (KDF S-70: Denken Co.). The final diameter after extrusion was 7 mm. The two types of extruded alloys were denoted as EXT923K and EXT973K. Microstructural observations were carried out by optical microscope (BX-51P: OLYMPUS) and SEM (JSM-6500F: JEOL). The microstructure of cast and extruded specimens was analyzed by SEM-EDS (EX-64175JMU: JEOL). For the electron back-scattered diffraction (EBSD) analysis, the specimens with dimension $10 \times 7 \times 3$ mm were cut parallel to the extrusion direction from extruded alloys. The specimens were ground with # 1000 and 4000 SiC abrasive papers, and then polished with 0.05 μm Al_2O_3 polishing suspension. Furthermore, they were electrochemical-polished in 30% HNO_3 in methanol at 243 K, and 30 V for 3 seconds. EBSD texture measurements were conducted by SEM, equipped with DigiView IV Detector (EDAX-TSL Co.) and OIM Data Collection 5.31 software (TSL Solutions K.K.). Tensile tests were carried out by using a universal testing machine (Autograph AG-X 50 kN: Shimadzu) with a strain rate of 5×10^{-4} /s. The extruded alloys were machined into tensile test specimens with 3 mm in diameter in accordance with JIS H 7405.

Table I. Chemical compositions of brass cast ingots. (mass%)

Alloys	Pb	Fe	Ti	Zn	Cu
Cu-40Zn	-0.005	0.003	---	38.40	Bal.
Cu-40Zn-0.5Ti	-0.005	0.004	0.496	38.96	Bal.

RESULTS AND DISCUSSION

Optical microstructures of Cu-40Zn and Cu-40Zn-0.5Ti are shown in Fig. 1. Both of the cast specimens consisted of α-β duplex phase structures. The α-β duplex phase of both specimens showed similar morphology. The grain size of cast specimens had approximately over 100 μm. Furthermore, Cu-40Zn-0.5Ti alloy specimens contained dispersoids of coarse IMCs having 10-30 μm in diameter. Moreover, the small amount of 0.09 ~ 0.16 at.% solute elemental Ti was detected on α-β phase in Cu-40Zn-0.5Ti cast alloy by SEM-EDS point analyses. Since the content of solute Ti element in α-β phase was quite small, almost all addition of elemental Ti existed as IMCs in cast ingots. Furthermore, since IMCs contained 24 at.% Ti, 49 at.% Cu, and 27 at.% Zn as the almost similar compositions of Cu_2TiZn in a previous research[15], IMCs in Cu-40Zn-0.5Ti alloy specimens were identified as Cu_2TiZn. Figure 2 shows optical microstructures of the cast specimens which were heat-treated at 973K for 15 min and following water-quenched. The area fraction of α-phase was decreased by diffusion and phase transformations from α-phase to β-phase in each sample, when the solid solution treatment was conducted at 973 K for 15 min. On the other hand, Cu_2TiZn IMCs never be observed in Cu-40Zn-0.5Ti which were solid solution treated at 973 K for 15 min. Since the content of solid solutionizing Ti element in the β-phase increased from 0.13 at.% to 0.66 at.% (acquired by SEM-EDS analysis), the IMCs were completely soluble in the matrix.

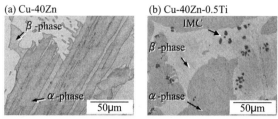

Figure 1. Optical microstructures of brass cast ingots: Cu-40Zn (a) and Cu-40Zn-0.5Ti(b).

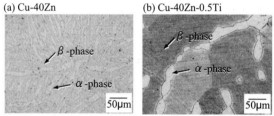

Figure 2. Optical microstructures of solid soluted specimens at 973 K for 15 min: Cu-40Zn (a) and Cu-40Zn-0.5Ti (b).

Optical microstructures of the extruded alloys denoted as EXT923K and EXT973K on the parallel cross-section of the extrusion direction are shown in Fig. 3. When cast alloys were extruded, Cu-40Zn[EXT923K] and Cu-40Zn-0.5Ti[EXT923K] were pre-heated the Cu-40Zn and Cu-40Zn-0.5Ti cast billet at 923 K for 3 min, respectively, and Cu-40Zn-0.5Ti[EXT973K] was pre-heated The Cu-40Zn-0.5Ti cast billet at 973 K for 15 min in solid solution treatment condition. The α-β duplex phase structures of the extruded specimens were slightly elongated along the extrusion direction in each extruded alloy. They were also finer than that of cast specimens because of the dynamic recrystallization of α-β phase. Their detail grain distribution will be mentioned later. Furthermore, the coarse Cu_2TiZn IMCs were dispersed in the matrix of Cu-40Zn-0.5Ti[EXT923K] as the same results of the observation in Cu-40Zn-0.5Ti cast specimens as shown in Fig. 1. A small amount of fine precipitation particles with 0.5 μm in diameter was also dispersed in the matrix of Cu-40Zn-0.5Ti[EXT923K]. On the other hand, instead of coarse Cu_2TiZn IMCs, the fine precipitation particles were densely distributed in the α-β duplex phase matrix of Cu-40Zn-0.5Ti[EXT973K] as shown in Fig. 3 (c). In the case of Cu-40Zn-0.5Ti[EXT973K], since coarse Cu_2TiZn IMCs were completely soluble in the matrix during pre-heating the billets before extrusion as shown in Fig. 2 (b), these fine particles in the matrix were precipitated from soluted Ti element in α-β phase during cooling after the extrusion.

Figure 4 shows the grain distribution of α-β duplex phase structures on EXT923K and EXT973K in the parallel cross-section of the extrusion direction acquired by EBSD analysis. Cu-40Zn-0.5Ti[EXT973K] had finer and more uniform α-β duplex phase structures compared with Cu-40Zn[EXT923K] and Cu-40Zn-0.5Ti[EXT923K]. It is considered that cast specimens were constituted by larger area fraction of β-phase at the extrusion condition of 973 K for 15 min compared with that of extrusion condition of 923 K for 3 min. This is due to the nucleation of fine α-phase from β-phase during cooling after the extrusion[16], and the nucleated grain growth was inhibited by the pinning effect of the fine precipitation particles at the grain boundaries.

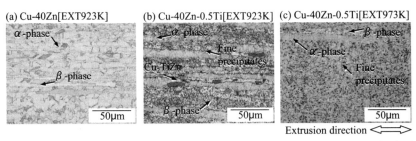

Figure 3. Optical microstructures of extruded specimens which were pre-heated at 923 K for 3 min: Cu-40Zn[EXT923K] (a) and Cu-40Zn-0.5Ti[EXT923K] (b), and at 973 K for 15 min: Cu-40Zn-0.5Ti[EXT973K] (c).

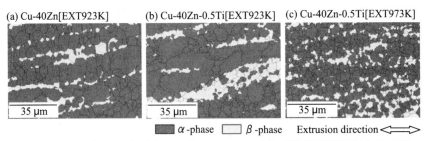

Figure 4. Grain distribution of α-β duplex phase structures of Cu-40Zn[EXT923K] (a), Cu-40Zn-0.5Ti[EXT923K] (b), and Cu-40Zn-0.5Ti[EXT973K] (c) by EBSD analysis.

Table 2 shows the characteristics of microstructures and mechanical properties in each extruded specimen. The area fraction and average grain size of extruded specimens in the parallel cross-section in the extrusion direction were acquired by EBSD analysis. Area fraction of α-phase and β-phase was indicated similar value in each extruded specimen. On the other hand, the grain size of α-β duplex phase in Cu-40Zn-0.5Ti[EXT973K] was 2.14 μm, 39 % finer than that in Cu-40Zn-0.5Ti[EXT923K] (3.51 μm). This is due to the density-distributed fine precipitation particles as above mentioned. Furthermore, yield strength (YS) and ultimate tensile strength (UTS) of Cu-40Zn-0.5Ti[EXT923K] were 226 MPa and 519 MPa, respectively as shown in Table 3. The YS of Cu-40Zn-0.5Ti[EXT923K] was comparable with that of extruded Cu-40Zn[EXT923K] (244 MPa) with no additives. This is because the extruded Cu-40Zn[EXT923K] had an average grain size of 3.89 μm and area fraction of α:β = 16:84, which were almost similar to that of Cu-40Zn[EXT923K] as shown in Table 3. YS of the extruded specimens was determined by the balance between grain size and area fraction of α and β-phase. On the other hand, YS and UTS of Cu-40Zn-0.5Ti[EXT973K] were 304 MPa and 543 MPa, 35 % and 4 % higher than that of Cu-40Zn-0.5Ti[EXT923K], respectively. Cu-40Zn-0.5Ti[EXT973K] had also 44% elongation. Cu-40Zn-0.5Ti[EXT973K] revealed excellent strength and ductility. Particularly, the increment of YS was due to the grain refinement strengthening. The grain refinement strengthening $\Delta\sigma_y$ was calculated by the Hall-Petch relationship

$$\Delta\sigma_y = k \times (d_{EXT973K}^{-1/2} - d_{EXT923K}^{-1/2})$$

where k is Hall-Petch coefficient: 528 N/μm$^{3/2}$ [17]; in the cast of Cu-40Zn-Ti alloy, $d_{EXT973K}$ and $d_{EXT923K}$ are average grain diameter 2.14 μm and 3.51 μm, respectively, as shown in Table 3. The calculated $\Delta\sigma_y$ was 79 MPa by the above Hall-Petch relationship. This value was similar to the experimental increment value of YS (78 MPa). Since Cu-40Zn-0.5Ti[EXT923K] and Cu-40Zn-0.5Ti[EXT973K] had also similar level of area fraction of between α and β-phase, mainly strengthening mechanism of extruded specimens in different thermo-mechanical treatment conditions was grain refinement strengthening.

SEM observations of fractured surface of tensile test specimen of Cu-40Zn-0.5Ti[EXT923K] and Cu-40Zn-0.5Ti[EXT973K] are shown in Fig. 5. Fine and uniform dimples were observed on fractured surface of Cu-40Zn-0.5Ti[EXT973K], due to the fine α-β duplex phase structures as previously mentioned. On the other hand, some coarse particles with brittle surfaces distributed in the center of large dimples were observed on fractured surface of Cu-40Zn-0.5Ti[EXT923K]. These brittle fractured surface areas were identified as Cu_2TiZn IMCs by SEM-EDS analysis. Such coarse Cu_2TiZn IMCs decreased the ductility of wrought brass materials. As a result, Cu-40Zn-0.5Ti[EXT973K] without coarse Cu_2TiZn IMCs exhibited the suitable elongation similar to that of Cu-40Zn-0.5Ti[EXT923K].

Table II. The characteristics of microstructure of Cu-40Zn[EXT923K], Cu-40Zn-0.5Ti[EXT923K], and Cu-40Zn-0.5Ti[EXT973K] were obtained by EBSD analysis and tensile test

	Area fraction (%)		Grain size /μm			UTS /MPa	YS /MPa	Elong. (%)
	α	B	α	β	α-β			
Cu-40Zn EXT923K	84	16	4.51	2.57	3.89	467	244	51
Cu-40Zn-Ti EXT923K	80	20	3.83	2.64	3.51	519	226	46
Cu-40Zn-Ti EXT973K	79	21	2.15	2.1	2.14	543	304	44

(a) Cu-40Zn-0.5Ti[EXT923K]   (b) Cu-40Zn-0.5Ti[EXT973K]

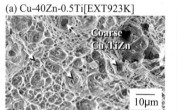 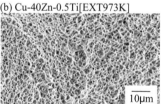

Figure 5. SEM observation of fractured surface of tensile test specimens of Cu-40Zn-0.5Ti[EXT923K] (a) and Cu-40Zn-0.5Ti[EXT973K] (b).

CONCLUSION

In this study, microstructural and mechanical properties of Cu-40Zn-0.5Ti ternary alloys which were solid solutionized and then extruded were investigated. The extruded specimens consisted of fine and uniform α-β duplex phase structures having 2.14 μm in diameter dispersed with fine precipitation particles having about 0.5 μm in diameter. Particularly, α-phase consisted of fine grains due to the pining effect by the fine precipitation at the grain boundaries. The tensile test results indicated YS: 304

MPa, UTS: 543 MPa and 44 % elongation. The high strengthening mechanism of the wrought brass alloy was mainly due to the grain refinement of α and β phases by the fine precipitates.

ACKNOWLEDGMENTS

Nippon atomized metal powders corporation is acknowledged for their help with preparing brass cast ingots.

REFERENCES

[1]Valerie Randle and Helen Davies, Evolution of Microstructure and Properties in Alpha-Brass after Iterative Processing, *Metall. and Mater. Trans. A*, **33**A, JUNE (2002) 1853-1856.

[2]G. Pantazopoulos, Leaded Brass Rods C 38500 for Automatic Machining Operations: A Technical Report, J. of Mater. *Eng. and Performance*, **11**, (4) August (2002) 402-407.

[3]J. X. Wu, M. R. Ji, M. Galeotti, A. M. Giusti and G. Rovida, Surface Composition of Machined Leaded Brass, *Surf. and Interface Anal.*, **22**, (1994) 323-326.

[4]Hisashi Imai, Yoshiharu Kosaka, Akimichi Kojima, Shufeng Li, Katsuyoshi Kondoh, Junko Umeda, Haruhiko Atsumi, Characteristics and machinability of lead-free P/M Cu60–Zn40 brass alloys dispersed with graphite, *Powder Technol.*, **198**, (2010) 417-421.

[5]Ulrika Borggren and Malin Selleby, A Thermodynamic Database for Special Brass, *J. of Phase Equilibria*, **24**, 2 2003.

[6]H. Mindivan, H. Çimeno˘glu and E.S. Kayali, Microstructures and wear properties of brass synchroniser rings, *Wear*, **254**, (2003) 532-537.

[7]M. Sundberg, R. Sundberg, S. Hogmark, R. Otterbergc , B. Lehtinenc, S. E. Hgrnstrgm and S. E. Karlsson, Metallographic Aspects on Wear of Special Brass, *Wear*, **115**, (1987) 151-165.

[8]G. Mauvoisin, O. Bartier, R. El Abdi, A. Nayebi, Influence of material properties on the drilling thrust to hardness ratio, Int. *J. of Machine Tools & Manufacture*, **43**, (2003) 825-832.

[9]C. Vilarinho, J.P. Davim, D. Soares, F. Castro, J. Barbosa, Influence of the chemical composition on the machinability of brasses, *J. of Mater. Process. Techno.*, **170**, (2005) 441-447.

[10]Lai-Zhe Jin and Rolf Sandstrm, Machinability data applied to materials selection, *Mater. & Design*, **15**, 6 (1994) 339-346.

[11]H. Atsumi, H. Imai, S. Li, Y. Kousaka, A. Kojima, K. Kondoh, Microstructure and Mechanical Properties of High Strength Brass Alloy with Some Elements, *Mater. Sci. Forum*, **654 – 656**, (2010) 2552-2555.

[12]S. Li, H. Imai, H. Atsumi and K. Kondoh, Characteristics of high strength extruded BS40CrFeSn alloy prepared by spark plasma sintering and hot pressing, *J. of Alloys and Compd.*, **493**, 18 (2010) 128-133.

[13]Shufeng Li, Hisashi Imai, Haruhiko Atsumi, Katsuyoshi Kondoh, Contribution of Ti addition to characteristics of extruded Cu40Zn brass alloy prepared by powder metallurgy, *Mater. & Design*, **32**, (2011) 192-197.

[14]Susumu Ikeno, Kenji Matusda, Youhei Nakamura, Tokimasa Kawabata, Yasuhiro Uetani, TEM Observation of α-phase in 60/40 Brass with Additional Element of Si, Mg or Ni, *J. of the JIRICu.*, **43**, 1 (2004).

[15]D. Soares, C. Vilarinho and F. Castr, Contribution to the knowledge of the Cu-Zn-Ti system for compositions close to brass alloys, *Scand. J. of Metall.*, **30**, (2001) 254–257.

[16]Carlo Mapelli and Roberto Venturini, Dependence of the mechanical properties of an a/b brass on the microstructural features induced by hot extrusion, *Scr. Mater.*, **54**, (2006) 1169–1173.

[17]Ewald Werner and Hein Peter Stijwe, Phase Boundaries as Obstacles to Dislocation Motion, *Mater. Sci. and Eng.*, **68**, (1984-1985) 175-182.

THE CHARACTERISTICS OF HIGH STRENGTH AND LEAD-FREE MACHINABLE α-β DUPLEX PHASE BRASS Cu-40Zn-Cr-Fe-Sn-Bi ALLOY

H. Atsumi[1], H. Imai[2], S. Li[2], K. Kondoh[2], Y. Kousaka[3], and A. Kojima[3]

[1] Graduate School of Engineering, Osaka University, 565-0871, 2-1, Yamadaoka, Suita, Osaka, Japan
[2] Joining and Welding Research Institute, Osaka University, 567-0047, 11-1, Mihogaoka, Ibaraki, Osaka, Japan
[3] San-etsu Metals Co. Ltd., 939-1315, 1892, Ohta, Tonami, Toyama, Japan

ABSTRACT

High strength and lead-free machinable α-β duplex phase brass Cu-40Zn-0.3Cr-0.2Fe-0.6Sn (mass%) alloys with 1-3 mass% Bi (Cu-40Zn-Cr-Fe-Sn-Bi) were prepared by casting process, and their microstructures, mechanical properties, and machinability were investigated. The extruded Cu-40Zn-Cr-Fe-Sn-Bi alloy consisted of fine and uniform α-β duplex phases dispersed with fine Bi particles. The number of their Bi particles was 1500-3000 /mm² in the transversal cross-section of the extrusion direction. Yield strength (YS) and ultimate tensile strength (UTS) of the extruded Cu-40Zn-Cr-Fe-Sn-Bi alloy were an average value of 288 MPa and 601 MPa, respectively. This extruded alloy revealed an increment of 29 % YS and 40 % UTS compared to the traditional machinable brass Cu-40Zn with 3.2 mass% Pb (Cu-40Zn-Pb). Since the machinability of the extruded Cu-40Zn-Cr-Fe-Sn-Bi alloy also maintained 75% of that of Cu-40Zn-Pb, they remarkably deviated from the traditional trade-off balance between hardness and machinability in the conventional machinable brass materials.

INTRODUCTION

Brass alloys (Cu-Zn) are widely applied for lead frames, connectors and other electronic components, pipes, valves, and so on, because of their excellent electrical and thermal conductivities, corrosion resistance, and good formability. In addition, 2-5 mass% lead (Pb) is added to the conventional brasses to improve their machinability[1-3]. However, Pb additives in the materials are significantly severe hazard to the environment and human health[4,5]. The material designs to safety the representative regulation as Restriction of Hazardous Substance (RoHS) and Waste Electrical and Electronic Equipment (WEEE) have been considered[6-8]. Bismuth (Bi) and graphite particles as alternative elements to Pb have been discussed to improve machinability of brass[9-11]. Bi has similar properties to Pb such as melting point, density, and solid solubility to copper. Furthermore, Bi causes no collateral effect to the environment and human health. On the other hand, the methods of solid solution strengthening and increasing area fraction of hard β-phases have been applied to brass materials for improvement of their mechanical properties. Cu-Zn binary phase diagram suggests that brass with 35-48 mass% Zn had α-β duplex phase structures to obtain suitable strength and hot forginability. For example, the commercial high-strength brass is invented and has been used for marine propellers because of its high mechanical properties and corrosion resistance. This material is alloyed with Al, Mn, and Fe for solid solution strengthening and increasing area fraction of hard β-phase[12,13]. However, when materials are added with the larger amounts of elements for strengthening, the coarser and brittle intermetallic compounds (IMCs) are produced in the matrix. In particular, such IMCs drastically decrease a machinability of brass materials. A trade-off balance between hardness and machinability was reported in the previous literatures[14,15]. In the present study, new high strength and lead-free machinable α-β duplex brass alloys dispersed with Bi particles were produced by casting process, and extruded consequently. The effects of Bi dispersoids on mechanical properties and machinability of the extruded specimens were investigated.

MATERIALS AND METHODS

Brass alloys were prepared by casting process. Cu-40Zn with additions of 0.3 mass% chromium (Cr), 0.2 mass% iron (Fe), and 0.6 mass% tin (Sn) (Cu-40Zn-Cr-Fe-Sn, CAST1) was prepared as a reference material, and Cu-40Zn-Cr-Fe-Sn with 1-3 mass% Bi (Cu-40Zn-Cr-Fe-Sn-Bi, CAST2-4) were alloyed at 1273-1473 K. The chemical compositions of cast ingots are shown in Table 1. The additions of elemental Cr, Fe and Sn strengthened α-β phase brass as small amounts of solid solution elements as informed in the previous reports[16,17]. The cast ingots with 60 mm in diameter were machined into billets with 41 mm in diameter. The billets were extruded by a 2000 kN hydraulic press machine (SHP-200-450: Shibayamakikai Co.). Before extrusion, the billets were preheated at 923 K for 180 seconds in Ar gas atmosphere in a muffle furnace (KDF S-70: Denken Co.). The final diameter after hot-extrusion was 7 mm. The extruded CAST1-4 were denoted as EXT1-4 in this study. The microstructural observation and analysis of the cast and extruded cast specimens were carried out by an optical microscope (OM, BX-51P: OLYMPUS) and SEM-EDS (JSM-6500F: JEOL, EX-64175JMU: JEOL). Back-scattered electron images (BSIs) acquired by SEM were used to evaluate the morphology of Bi particles dispersed in the extruded cast specimens. The number of Bi particles was measured by the image analyses of BSIs. For the electron back-scattered diffraction (EBSD) analyses, the specimens with dimension 10×7×3 mm were cut parallel to the extrusion direction from the extruded specimens. They were ground with # 1000 and 4000 SiC abrasive papers, and then polished with 0.05 μm Al_2O_3 polishing suspension. Furthermore, they were electrochemical-polished in 30% HNO_3 methanol solution at 243 K for 3 seconds, using a DC power supply with 30 V. EBSD texture measurements were conducted by SEM equipped with DigiView IV Detector (EDAX-TSL Co.) and OIM Data Collection 5.31 software (TSL Solutions K.K.). For machinability evaluation, drilling test was carried out by using a drill tool (EX-SUS-GDS: OSG Co.), having a 4.5 mm diameter, under dry conditions. The drill rotation speed was 900 rpm, and applied load was 9.8 N during drilling. The drilling time to make a hole with a 5 mm depth was measured. After continuously repeating this drilling test 10 times, the average drilling speed was used as a machinability parameter of the specimens. Mechanical properties were evaluated by tensile tests and hardness tests. The extruded specimens were machined into tensile test specimens with 3 mm diameter in accordance with JIS H 7405. Tensile tests were carried out on a universal testing machine (Autograph AG-X 50kN: Shimadzu) with a strain rate of $5×10^{-4}$ /s. Hardness tests were performed by a Vickers micro hardness tester (HMV-2T: Shimadzu) with testing load 245.2 mN (0.025 kgf) for 15 seconds at room temperature.

Table I. Chemical compositions of brass cast ingots used in this study. (mass%)

Alloys	Sn	Pb	Zn	Fe	Cr	Bi	Cu
CAST1	0.590	-0.005	40.86	0.220	0.340	---	Bal.
CAST2	0.595	-0.005	40.81	0.229	0.256	0.994	Bal.
CAST3	0.600	-0.005	40.64	0.230	0.260	2.020	Bal.
CAST4	0.578	-0.005	40.83	0.219	0.220	2.850	Bal.

RESULTS AND DISCUSSION

Microstructural and mechanical properties of extruded alloys

Optical microstructures of extruded specimens (EXT1-4) are shown in Fig. 1. EXT1 without Bi had α-β duplex phase structures with discrete Cr-Fe IMCs along the extrusion direction, which were mechanically broken and formed fine particles by hot-extrusion. The increasing amount of elemental Bi in EXT2-4 caused an increasing area fraction of β-phase with decreasing the fraction of Cu content to Zn content. This is because the Zn content was kept to 40 mass% in cast ingots as shown in Table 1. In addition, each extruded specimen showed fine α-phase grains. It is considered that cast alloys were constituted by the larger area fraction of β-phase at the extrusion condition of 923 K and 180 seconds holding time, comparing with that of room temperature as can be expected from Cu-Zn binary phase diagram. This is due to the nucleation of the α-phase from β-phase during cooling after hot-extrusion [18]. Bi particles were also slightly elongated along the extrusion direction. Their detail morphology will be mentioned later. The distributions of α-β duplex phase structures on the transversal cross-section of EXT1-4 in the extrusion direction by EBSD are shown in Fig. 2. The area fraction of β-phase in EXT1-4 was 60 %, 57 %, 62 %, and 80 %, respectively. The average grain size is shown in right bottom corner of each image. EXT4 with the large area fraction of β-phase consisted of coarser grains compared with EXT1-3 specimens. It is considered that the grain size of extruded α-β brass was determined by the phase transformation from β-phase to α-phase. Extruded α-β brass had the fine structures due to the nucleation of fine α-phase during cooling after hot-extrusion as above mentioned. Therefore, since EXT4 had the high concentration of elemental Zn, it had the coarse structures without much phase transformation from β-phase to α-phase. Figure 3 shows a dependence of the tensile properties on the Bi content of their extruded alloys. The extruded high-strength brass alloy Cu-40Zn-Cr-Fe-Sn (EXT1) had yield strength (YS): 285 MPa and ultimate tensile strength (UTS): 582 MPa, which were, respectively, 17 % and 25 % higher than extruded Cu-40Zn had (YS: 244 MPa and UTS: 467 MPa). EXT1 was strengthened by both of the solid solution elements of Cr, Fe and Sn and the increasing area fraction of hard β-phase. YS and UTS of the extruded Cu-40Zn-Cr-Fe-Sn-Bi (EXT2-4) had an average value of 288 MPa and 601 MPa, as the same level of that of EXT1, respectively, due to the same strengthening mechanisms. Besides, the tensile properties of EXT2-4 were determined by a balance between the grain size and the area fraction of β-phase. On the other hand, elongation of EXT2-4 decreased with increasing the content of elemental Bi addition. In particular, the elongation of EXT4 (2.85 mass% Bi) was indicated 22 %, as the lowest value in the extruded specimens. Figure 4 shows the SEM image and the SEM-EDS map for Bi elements in the fractured surface of the tensile specimens of EXT4. Some coarse particles with brittle surfaces distributing in the cracks were observed on fractured surface of EXT4. These brittle fractured surface areas were identified as Bi particles by SEM-EDS analysis. Since Bi particles were soft but brittle materials, the elongation of extruded specimens was decreased these coarse Bi particles.

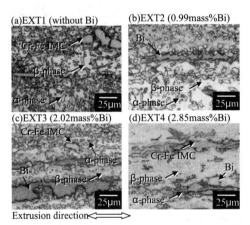

Figure 1. Optical microstructures of extruded cast brass specimens without Bi (a), with 0.99 mass% Bi (b), with 2.02 mass% Bi (c), and with 2.85 mass% Bi (d).

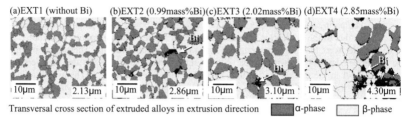

Figure 2. Grain distribution of α-β duplex phase structures on EXT1 (a), EXT2 (b), EXT3 (c), and EXT4 (d) acquired by EBSD analysis.

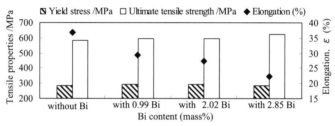

Figure 3. Tensile properties of extruded cast brassspecimens without Bi, with 0.99 mass% Bi, with 2.02 mass% Bi, and with 2.85 mass% Bi.

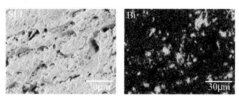

Figure 4. SEM-EDS observation on fractured surface of tensile test specimens of EXT4 (2.85 mass% Bi).

Relationship between morphology of Bi particles and machinability of extruded specimens

Figure 5 shows BSIs of the transversal cross section of EXT2-4 in the extrusion direction. The BSI indicates the binarized image which segregated Bi particles from α-β phase matrix. Since elemental Bi was heavier than elemental Cu, Zn, Cr, Fe and Sn, white areas indicated the Bi particles in the matrix by the dependence of atomic number. The number of Bi particles in a certain analysis area was measured by the image analyses of BSIs. Table 2 shows the Bi particle density and the drilling speed, and summarized the mechanical properties in each extruded alloy. The drilling speed of extruded specimens was increased in proportion to the number of Bi particles in the matrix. There were a good correlation between machinability and the number density of Bi particles in the matrix. However, EXT4 (with 2.85mass% Bi) did not have enough machinability comparing with EXT2 and EXT3. Though EXT4 had good dispersion of Bi elements in its matrix, it had coarser grains compared with other specimens. A previous research reported that the alloys with coarse grains had poor machinability, because of the sliding wear and friction resistance between the cutting tools and material surface [19]. Figure 6 shows a relationship between tensile property (UTS) and machinability (drilling speed) of the extruded cast brass materials. As references, data of extruded Cu-40Zn-X%Bi (X = 1.0, 2.33, 2.53, 2.6, 2.95, and 5.45 mass%) and conventional machinable brass Cu-40Zn with 3.2 mass% Pb (Cu-40Zn-Pb) were also described. When extruded alloys contained coarse Bi dispersoids, the tensile property of extruded Cu-40Zn-X%Bi decreased with increasing the machinability according to a trade-off balance between hardness and machinability. The extruded Cu-40Zn-Cr-Fe-Sn-Bi had high tensile properties and an excellent machinability, because it had the better dispersion of Bi particles than the extruded Cu-40Zn-X%Bi. In addition, the UTS of extruded Cu-40Zn-Cr-Fe-Sn-Bi had 40 % higher than that of the conventional extruded Cu-40Zn-Pb. Since its drilling speed maintained 75 % of machinability of extruded Cu-40Zn-Pb, extruded Cu-40Zn-Cr-Fe-Sn-Bi deviated from the traditional trade-off balance between tensile strength and machinability in the conventional machinable brass materials.

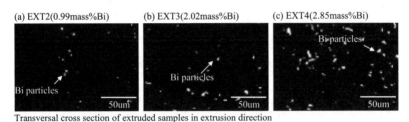

Figure 5. Back scattered electron images of EXT2(0.99mass% Bi) (a), EXT3(2.02mass% Bi) (b), and EXT4(2.85mass% Bi) (c) by back-scattered electron images.

Table II. The characteristics of microstructural and mechanical properties of each extruded specimen.

Specimen	Bi content	Grain size /μm		Bi particle density	Drilling speed	UTS	Elong.	Hardness
	mass%	α-phase	β-phase	/mm²	mm/sec	MPa	%	Hv
EXT1	0	1.96	3.47	---	---	582	37	154±4
EXT2	0.994	2.51	4.05	940	0.14	597	29	157±8
EXT3	2.02	2.83	3.87	1834	0.34	596	27	155±7
EXT4	2.85	3.98	5.52	3221	0.28	622	22	159±7

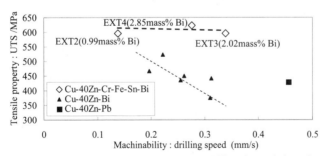

Figure 6. Relationship between tensile property and machinability of extruded cast brass materials.

CONCLUSION

In this study, microstructures, mechanical properties, and machinability of high strength and lead-free machinable α-β duplex phase brass Cu-40Zn-Cr-Fe-Sn-Bi were investigated. The extruded Cu-40Zn-Cr-Fe-Sn-Bi consisted of fine α-β phases containing the fine and uniform Cr-Fe IMCs and Bi particles. YS and UTS of the extruded Cu-40Zn-Cr-Fe-Sn-Bi were in average 288 MPa and 601 MPa, respectively. They were 29 % and 40 % higher than YS and UTS of the conventional machinable brass Cu-40Zn-Pb, respectively. Since its machinability also maintained 75% of that of Cu-40Zn-Pb, the extruded Cu-40Zn-Cr-Fe-Sn-Bi deviated from the traditional trade-off balance between hardness and machinability in the conventional machinable brass materials.

ACKNOWLEDGMENTS

Nippon atomized metal powders corporation is acknowledged for their help with preparing brass cast ingots used in this study.

REFERENCES
[1]G. Pantazopoulos, Leaded brass rods C 38500 for automatic machining operations: A technical report, *J. of Mater. Eng. and Perform.*, **11**, (2002) 402-407.
[2]Japan Capper and Brass Association, *Base and Industrial technology of copper and copper alloys* (in Japanese) (1994).
[3]P. García, S. Rivera, M. Palacios, and J. Belzunce, Comparative study of the parameters influencing the machinability of leaded brasses, *Eng. Fail. Anal.*, **17**, (2010) 771–776.
[4]Xi. Chen, Anmin Hu, Ming Li, and Dali Mao, Study on the properties of Sn–9Zn–xCr lead-free solder, *J. Alloys Compd.*, **460**, (2008) 478–484.
[5]S. Kuyucak, and M. Sahoo, A review of the machinability of copper-base alloys, *Can. Metall. Q.*, **35**, (1) (1996) 1–15.
[6]Bob D'Mellow, David J. Thomas, Malcolm J. Joyce, Peter Kolkowski, Neil J. Roberts, and Stephen D. Monk, The replacement of cadmium as a thermal neutron filter, *Nucl. Instrum. Methods Phys. Res. A*, **577**, (3) (2007) 690–695.
[7]RoHS and WEEE compliant flame retardants developed for electrical connectors, *Reinf. Plastics*, **51**, (8) (2007) 12.
[8]Christian Mans, Stephanie Hanning, Christoph Simons, Anne Wegner, Anton Janβen, and Martin Kreyenschmidt, Development of suitable plastic standards for X-ray fluorescence analysis, Spectrochim. *Acta Part B*, **62**, (2007) 116–122.
[9]Whiting, L V, Newcombe, P D, and Sahoo, M, Casting characteristics of red brass containing bismuth and selenium, *Trans. of the Am. Foundrymen's Soc.*, **103**, (1995) 683-691.
[10]A. La Fontaine, and V.J. Keast, Compositional distributions in classical and lead-free brasses, *Mater. Charact.*, **57**, (2006) 424–429.
[11]H. Imai, Y. Kosaka, A. Kojima, S. Li, K. Kondoh, J. Umeda, and H. Atsumi, Develpment of Lead-free Machinable Brass with Bismuth and Grapite Particles by Powder Metallurgy Process, *Mater. Trans.*, **51**, (2010) 855-859.
[12]H. Mindivan, H. Çimeno˙glu, and E.S. Kayali, Microstructures and wear properties of brass synchroniser rings, *Wear*, **254**, (2003) 532–537.
[13]M. sundberg, R. sundberg, S. Hogmark, R. Otterbergc , B. Lehtinenc, S. E. Hgrnstrgm and S. E. Karlsson, Metallographic aspects on wear of special brass, *Wear*, **115**, (1987) 151 – 165.
[14]C. Vilarinho, J.P. Davim, D. Soares, F. Castro, and J. Barbosa, Influence of the chemical composition on the machinability of brasses, *J. Mater. Process Technol.*, **170**, (2005) 441–447.
[15]Lai-Zhe Jin and Rolf Sandström, Machinability data applied to materials selection, *Mater. & Design*, **15**, (1994) 339-346.

[16]H. Atsumi, H. Imai, S. Li, Y. Kousaka, A. Kojima, and K. Kondoh, Microstructure and Mechanical Properties of High Strength Brass Alloy with Some Elements, *Mater. Sci. Forum,* **654 – 656**, (2010) 2552-2555.

[17]S. Li, H. Imai, H. Atsumi and K. Kondoh, Characteristics of high strength extruded BS40CrFeSn alloy prepared by spark plasma sintering and hot pressing, *J. of Alloy. and Compd.*, **493**, (2010) 128-133.

[18]Carlo Mapelli and Roberto Venturini, Dependence of the mechanical properties of an a/b brass on the microstructural features induced by hot extrusion, *Scr. Mater.*, **54**, (2006) 1169–1173.

[19]Tamer El-Raghy, Peter Blau, and Michel W. Barsoum, Effect of grain size on friction and wear behavior of Ti$_3$SiC$_2$, *Wear*, **238**, (2000) 125–130.

PREPARATION OF BIOMASS CHAR FOR IRONMAKING AND ITS REACTIVITY

Hu Zhengwen, Zhang Jianliang, Zhang Xu, Fan Zhengyun, Li Jing

School of Metallurgical and Ecological Engineering, University of Science and Technology Beijing

Beijing 100083, China

ABSTRACT

Due to a large consumption of coal and coke, ironmaking process gives off huge quantities of waste gas into the air, such as carbon dioxide and sulfur dioxide. Therefore, biomass is supposed to be used in ironmaking process to partially replace metallurgical coal and coke for energy-saving and emission-cutting. As raw biomass is not suitable for ironmaking, pretreatment of biomass is needed. Laboratorial preparation of biomass char, suitable for industrial application, is carried out and proper preparation conditions of biochar are presented. And then, properties of biochar, especially its reactivity, are also investigated. It was concluded that biomass char is a kind of solid fuel with high reactivity and high purity, applicable to substitute for fossil fuels in ironmaking.

INTRODUCTION

A large quantity of carbon resources, coal and coke, are consumed in ironmaking process to supply heat and reducing agent, causing serious environmental problems and resource crisis. In conventional ironmaking process, about 500 kg of fossil carbon would be consumed to produce one ton of steel, and nearly 2 tons of CO_2 and other emissions would be emitted into the air simultaneously[1]. Biomass, carbon rich and carbon neutral, therefore, is getting more and more attention as a kind of renewable and clean energy in ironmaking process [2-3]. For example, Ueda S. et al had tried to add new burden with biomass char into the blast furnace to achieve low reducing agent operation and mitigate CO_2 emission [4-5]. Besides, some researchers attempted to utilizing biomass in coke making, iron ore sintering and blast furnace auxiliary injection [6-8].

Theoretically, all of the energy used in carbon-based ironmaking could be replaced by biomass carbon except the essential coke for large blast furnaces [9-10]. But in practice only a few steel plants use biomass to produce hot metal (for example, there are small blast furnaces in Brazil which use charcoal as reducing agent). The main reason is that raw biomass is not suitable for direct use in industry due to its high moisture content, poor grindability, low energy density and low bulk density, and consequently the pretreatment of raw biomass is indispensable[11].

As a kind of attractive renewable energy with a promising prospect, the carbonization and pyrolysis of lots of raw biomass materials were studied by many researchers around the world with a long history [12-13]. However, previous research about the utilization of biomass is almost centered on its combustion, gasification and pyrolysis for power generation or chemicals' production, and little information is about preparing biomass char for ironmaking. In this article, the optimum condition (heating pattern, carbonization temperature and holding time) for producing biochar from biomass was suggested through a series of carbonizing experiments. Furthermore, effect of carbonizing condition of biomass on the reactivity of derived biochar with CO_2 was investigated.

EXPERIMENTAL APPARATUS AND METHODS

1 Biomass material

A kind of woody biomass (white pine) has been used in this study. The white pine material, produced in the northeast of China, was obtained from the waste timber of a wood processing mill. Then the material was processed into rectangular woodblocks, 80×30×20mm, and some white pine sawdust (<150μm) was also prepared. The proximate and ultimate analyses of the biomass material are given in Table 1.

Table I.   Analyses of biomass sample (air dry basis)

Proximate analysis	wt,%	Ultimate analysis	wt,%
Moisture	4.98	Carbon	48.04
Volatile matter	78.18	Hydrogen	5.60
Fixed carbon	16.39	Oxygen	39.77
Ash	0.45	Nitrogen	0.37
		Total sulfur	0.06

2 Apparatus and methods

2.1 Carbonizing experiment

The carbonization of biomass samples was conducted in an electric resistance tubular furnace (12kw, inner heating chamber Φ120×250 mm), which was connected to a PID controller. Biomass samples were put in a Pt basket, which was hung on the inner wall of the carbonizing tube reactor. When the experiment started, the whole reactor, including a thermocouple suspended in it near the sample surface to detect its temperature, was put into the heating chamber. The schematic sketch of the carbonizing experiment is shown in Fig. 1.

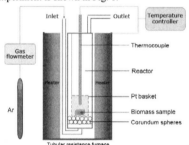

Figure 1. Schematic sketch of carbonizing experiment

Three key factors, namely carbonizing temperature, heating pattern and holding time, that affect the preparation of biomass char were considered. The biomass samples were carbonized at 300℃, 400℃, 500℃, 600℃ and 700℃, with a holding time of 30min, 60min and 90min respectively. As to the heating pattern, a room temperature heating pattern (the reactor was put into the heating chamber previously and heated from room temperature to the carbonizing temperature with the furnace, hereinafter called "RT" heating pattern) and a constant temperature heating pattern (the furnace was previously heated to the final carbonizing temperature and kept at the temperature, then put the reactor in it, hereinafter called "CT" heating pattern) were chosen. Biomass samples were heated to the final temperature by different heating pattern and kept for a certain time. Then the reactor was taken out of the furnace to cool to room temperature. The Ar gas was introduced into the carbonizing reactor at a flow rate of 5 L/min throughout the experiment to prevent biomass samples from burning.

2.2 Thermal gravimetric experiments

The combustibility and gasification reactivity of derived biomass char and the carbonizing process of sawdust were studied by a thermal gravimetric balance (produced by Beijing Optical Instrument Factory). The biochar samples were ground into particles less than 74μm, and for every thermal gravimetric experiment 16.5 mg of pulverized biochar sample was accurately weighed. The heating rate of the thermal gravimetric balance was set at 15℃/min, and finally the TG-DTG curves of samples would be recorded by a computer. As far as the carbonization test of sawdust is concerned, the Ar gas was introduced at a flow rate of 60mL/min until the end (1100℃); when it comes to the gasification experiment of biochar, when the sample was heated to 900℃, the protective Ar gas (60mL/min) was switched to CO_2 (60mL/min) until 1200℃. As to the combustibility test of biochar, the air(60mL/min) was introduced from room temperature to 900℃.

RESULTS AND DISCUSSION

1 Carbonizing process of raw biomass

The carbonization of raw biomass mainly includes the evaporation of moisture, the volatilization of volatile matter, and the decomposition of cellulose and lignin. The solid residue after carbonization is generally regarded as biomass char or biochar. Therefore, biochar is, in fact, mainly composed of carbon, residual volatile matter and ash. The TG curve of biochar (CT heating pattern, carbonized at 700℃ for 30min) and TG-DTG curves of raw biomass in Ar gas are shown in Fig. 2.

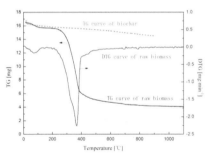

Figure 2. TG-DTG curves of biomass samples carbonized in Ar

It can be seen from TG-DTG curves of raw biomass that the carbonization process approximately has the following three stages:

1) The dehydration stage (about 50-110℃), corresponding to the first obvious trough of the DTG curve, was the beginning of the carbonization, accompanied by the loss of moisture.

2) The rapid carbonization stage (about 230-500℃) corresponding to the largest valley of the DTG curve. The thermal decomposition of hemicelluloses, cellulose and lignin, as well as the generation of combustible gas and solid biochar all happened in this stage. The maximum weight loss rate of biomass in this stage was about 2.19 mg/min, and the total weight loss was as high as 61.78%.

3) The carbonization-ending stage (over 500℃) was progressing much slowly with a weight loss rate less than 0.10 mg/min. The total weight loss from 500℃ to 700℃ is about 3.95%, and from 700℃ to 1200℃ was about 2.24%. A small amount of residual volatile matter in the sample would continue to separate out gradually in this stage.

The TG curve of biochar shows that there is hardly any volatile matter in the biochar (CT heating pattern, carbonized at 700℃ for 30min), and a slightly weight change occurred only when it was over 700℃.

Altogether, the main carbonizing reactions of biomass that affect its weight occurred at the temperature range of 230-500℃, basically completing at about 700℃, and further increase of temperature had little influence on the weight change.

2 The yield of biochar

Effect of carbonizing temperature, heating pattern and holding time on the yield of biochar is shown in Fig. 3. Where, symbols named "RT-30", "RT-60" and "RT-90" (with solid symbols) denote RT heating pattern and carbonized for 30min, 60min and 90min. Similarly, symbols named "CT-30", "CT-60" and "CT-90" (with hollow symbols) denote RT heating pattern and carbonized for 30min, 60min and 90min. In addition, the relationship between carbonizing temperature and instantaneous yield of biochar obtained by the thermo balance is denoted by solid pentagram.

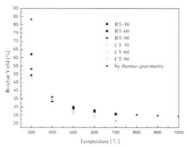

Figure 3.   Effect of carbonization conditions on the yield of biochar

2.1 Effect of carbonizing temperature on biochar yield

Carbonizing temperature is the most critical factor that affects the yield of biochar. As indicated in Figure 3, the yield of biochar samples all decreased gradually with the increase of carbonizing temperature. The yield fell rapidly when the biomass was carbonized at 300-400℃, and slowed down at the carbonizing temperature of 400-500℃. From 500 to 700℃, the yield reduced even more slowly, approximately liner decline. When it was over 700℃, the carbonizing temperature had hardly any effect on biochar yield. In view of the grindability and yield of biochar, the rational carbonizing temperature ranges from 500℃ to 700℃.

2.2 Effect of holding time on biochar yield

The biochar yield declined with the increment of holding time at the temperature range of 300-400℃. The main reason is that more volatile matter would separate out from the sample with a longer holding time. From 400 to 500℃, there is only a slightly influence of holding time on biochar yield, and no obvious effect of time on the yield would be observed over 500℃.

Consequently, when the carbonizing temperature is chosen at 500-700℃, the holding time for preparing biochar could be 30min or even shorter than that to improve production efficiency.

2.3 Effect of heating pattern on biochar yield

There is only a slightly effect of heating pattern on biochar yield when the carbonizing temperature is below 500℃. However, the effect becomes more and more obvious over 500℃. The biochar yield by RT heating pattern is a little higher than that by CT heating pattern. About 4.33% difference of average yield between RT and CT heating patterns is calculated at 700℃. Therefore, the RT heating pattern should be chosen for a slightly higher biochar yield; nevertheless, the CT heating pattern, not requiring cooling and heating from room temperature repeatedly, should be chosen for the continuous preparation of biochar in practice to reduce energy consumption and improve efficiency.

Consequently, the rational preparing process for biochar is to carbonize biomass material at about 500-700℃ for 30 minutes, considering the biochar yield, production efficiency and energy consumption.

3 Composition and microstructure of biochar

The proximate and ultimate analyses of the derived biochar sample (CT-500-30) are shown in Table 2. Although the ash content of biomass material increased after carbonization, it is still much lower than that of coal. Besides, biochar has a much more carbon content and much lower oxygen content than raw biomass. Compared with coal, biochar has comparable composition, much lower sulfur content and ash, which is a kind of promising energy for the substitution of fossil fuels.

Table II. Analyses of biochar sample (air dry basis)

Proximate analysis	wt,%	Ultimate analysis	wt,%
Moisture	1.40	Carbon	85.15
Volatile matter	16.63	Hydrogen	3.03
Fixed carbon	81.03	Oxygen	9.28
Ash	0.94	Nitrogen	0.25
		Total sulfur	0.05

The SEM pictures of biochar (cross section), pulverized biochar and comparative coal samples are shown in Fig. 4. It indicates that biochar has typical honeycombed duct structure with thin walls (approximately 1-5μm). Due to the decomposition of cellulose and lignin, biochar can be easily ground to particles less than 50μm. Even after grind, the pulverized biochar has different appearance from coal fine. One has plate-like shapes, while the other one is granular. Probably that is why biochar has a higher specific surface area, ensuring its good reactivity.

Figure 4. Microstructure of biochar and comparative coal samples

4 Reactivity of biochar

The carbon conversion rate, denoted by X_c, is introduced to evaluate the reactivity of biochar samples with CO_2. X_c is calculated by the following formula,

$$X_c = \frac{m_0 - m_T}{m_0 - m_\infty} \times 100\% \qquad (1)$$

Where m_0, m_T and m_∞ denotes the mass of samples before reaction, at the temperature of T and after reaction respectively. All of the above values can be obtained from TG curves of the gasification reaction.

4.1 Effect of carbonizing temperature on biochar's reactivity

The carbon conversion rates of biochar, RT-30min and CT-30min, react with CO_2 at 900-1100℃ are shown in Fig. 5. As a result of the instability of TG curves at 900-950℃, owing to the switching of Ar to CO_2 gas, there is little comparison among the carbon conversion rates. Therefore, the comparison of X_c is only conducted over 950℃. As for the RT heating pattern, X_c of the biochar carbonized at 600℃ (RT-600) is the lowest at the same gasification temperature, and the X_c of biochar samples carbonized at 500℃ and 700℃ are nearly as high as the same. As for the CT heating pattern, the X_c of biochar carbonized at 700℃ is the highest, then the CT-500 biochar, and the biochar carbonized at 400℃ is the lowest. As a consequence of the above, it is not really that biochar carbonized at a higher temperature (400-700℃) has better reactivity.

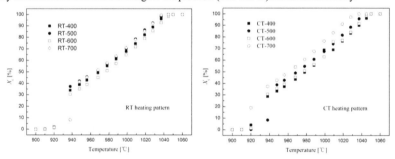

Figure 5.   Effect of carbonizing temperature on carbon conversion rate

4.2 Effect of holding time on biochar's reactivity

The X_c of biochar samples carbonized at 500℃ by CT heating pattern with different holding time are shown in Fig. 6. The X_c of "CT-500-90" is the lowest and that of "CT-500-30" is the highest. It can be concluded that the reactivity of biochar decreases with the extension of holding time. The possible reason is still that more volatile matter, affecting the reactivity directly, would separate out from the sample with a longer holding time. Therefore, the holding time for preparing biochar with good reactivity should be less than 30 minutes.

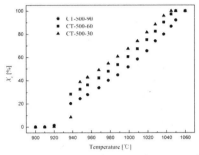

Figure 6.   Effect of holding time on carbon conversion rate

4.3 Effect of heating pattern on biochar's reactivity

The X_c of biochar samples carbonized at different temperatures by RT(with circle symbols) and CT(with square symbols) heating patterns for 30min are shown in Fig. 7. As for the biochar samples carbonized at 400°C, the "RT" is higher than the "CT"; as for 500°C, the "RT" is still higher than the "CT", but the difference between them diminishes; as for 600°C, there is nearly no difference between the "RT" and "CT", indicating almost the same reactivity; as for 700°C, the "CT" is higher than the "RT". These results suggest that when the carbonizing temperature is below 600°C, biochar samples carbonized by RT heating pattern have a better reactivity; when the carbonizing temperature is over 600°C, biochar samples carbonized by CT heating pattern have a better reactivity.

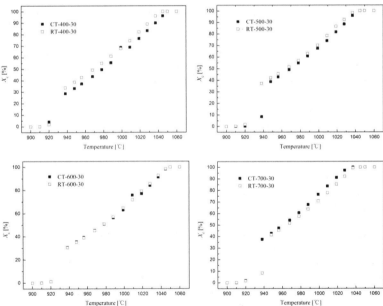

Figure 7.   Effect of heating pattern on carbon conversion rate

5 Preparing conditions of biochar for ironmaking

The yield and properties of biochar vary with the change of carbonizing condition. If only the yield and carbonizing process of biochar were considered, the carbonizing temperature could be at 400-700℃ by any heating pattern, and the holding time should be 30min, achieving a biochar yield of 20-35%. From an energy use and efficiency perspective, a lower carbonizing temperature should be chosen by CT heating pattern, and the holding time should also be 30min, realizing a shorter production cycle and higher energy efficiency. In terms of ironmaking process, requiring biochar with good reactivity simultaneously, the optimum condition is that biomass material be carbonized at 500 by CT heating pattern with 30min or less time, namely the "CT-500-30" process. The biochar prepared by "CT-500-30" has separated out most of the volatile matter, possessed of good reactivity and comparable composition to coal, and was sufficient to be used in ironmaking process as heating agent and reducing agent.

CONCLUSION

The rational preparing condition of biochar was investigated thoroughly. The composition, microstructure and gasification reactivity of derived biochar samples were also studied. Results can be summarized as follows.

(1) The main carbonizing process of biomass material occurred at about 230-500℃. The yield of biochar declined with the rising of carbonizing temperature. The holding time had no significant impact on the yield when biomass was carbonized at a temperature over 500℃. The content of volatile matter in the biochar carbonized at 700℃ was low, and a higher carbonizing temperature had trifle influence on the yield.

(2) Compared with raw biomass, biochar had extremely high fixed carbon content, very low volatile matter content and very good grindability, more suitable for industrial application. By comparison with coal, biochar had extremely low ash, low sulfur, unique microstructure and good reactivity, as suitable as coal for ironmaking process.

(3) The reactivity of biochar with CO_2 gradually deteriorated with the extension of holding time, but did not keep improving along with the rising of carbonizing temperature. In addition, the biochar carbonized by RT heating pattern had better reactivity than that by CT heating pattern below 600℃; while over 600℃, the case is contrary.

(4) The optimum process for preparing biochar is to carbonize biomass at 500℃ by CT heating pattern with a holding time of 30 minutes.

REFERENCES

[1]C. B. Xu, and D. Q. Cang, A brief overview of low CO_2 emission technologies for iron and steel making, *Journal of Iron and Steel Research International*, **17**(3), 1-7(2010).

[2] P. Mckendry, Energy production from biomass (part 1): overview of biomass, *Bioresource Technology*, **83**(1), 37-46(2002).

[3]J. Södermana, H. Saxéna, and F. Pettersson. Future potential for biomass use in blast furnace ironmaking, *Computer Aided Chemical Engineering*, **26**, 567-571(2009).

[4]S. Ueda, K. Watanabe, K. Yanagiya, et al. Improvement of reactivity of carbon iron ore composite with biomass char for blast furnace, *ISIJ International*, **49**(10), 1505-1512(2009).

[5]S. Ueda, K. Yanagiya, K. Watanabe, et al. Reaction Model and Reduction Behavior of Carbon Iron Ore Composite in Blast Furnace, *ISIJ International*, **49**(6), 827-836(2009).

[6]M. Zandi, M. Martinez-Pacheco, and T. A. T. Fray, Biomass for iron ore sintering, *Minerals*

Engineering, **23**, 1139-1145(2010).

[7]A. Babich, D. Senk, and M. Fernandez, Charcoal Behaviour by Its Injection into the Modern Blast Furnace, *ISIJ International*, **50**(1), 81-83(2010).

[8]T. Matsumura, M. Ichida, T. Nagasaka, et al. Carbonization behaviour of woody biomass and resulting metallurgical coke properties, *ISIJ International*, **48**(5), 572-577(2008).

[9]Z. W. Hu, J. L. Zhang, H. B. Zuo, et al. Substitution of Biomass for Coal and Coke in Ironmaking Process, *Advanced Materials Research*, **236**, 77-82(2011).

[10]H. Helle, M. Helle, H. Saxéna, et al. Mathematical Optimization of Ironmaking with Biomass as Auxiliary Reductant in the Blast Furnace, *ISIJ International*, **49**(9), 1316-1324(2009).

[11]H. Abdullah, and H. Wu, Biochar as a Fuel: 1. Properties and Grindability of Biochars Produced from the Pyrolysis of Mallee Wood under Slow-Heating Conditions, *Energy & Fuels*, **23**(8), 4174-4181(2009).

[12]P. Mckendry, Energy production from biomass (part 2): conversion technologies, *Bioresource Technology*, **83**(1), 47-54(2002).

[13]W. J. Desisto, N. Hill, S. H. Beis, et al. Fast Pyrolysis of Pine Sawdust in a Fluidized-Bed Reactor, *Energy & Fuels*, **24**(4), 2642-2651(2010).

INTELLIGENT ENERGY SAVING SYSTEM IN HOT STRIP MILL

H. Imanari [1)], K. Ohara [1)], K. Kitagoh [1)], Y. Sakiyama [2)], F. Williams [3)]

1) Research and Development Center of Process Control, Toshiba Mitsubishi-Electric Industrial Systems Corporation (TMEIC), Minato-ku, Tokyo, Japan
2) Steel Making, Instrumentation & Energy Saving Metals Process Systems Engineering Department, Toshiba Mitsubishi-Electric Industrial Systems Corporation (TMEIC), Minato-ku, Tokyo, Japan
3) Process Automation Systems Engineering, Toshiba Mitsubishi-Electric Industrial Systems Corporation (TMEIC), Roanoke, Virginia, USA

ABSTRACT

Reducing energy usage in steel industries has a large impact on creating a more sustainable society. Toshiba Mitsubishi-Electric Industrial Systems Corporation, TMEIC, as an electrical equipment supplier to iron and steel industries, is developing new energy saving solutions for hot strip mill lines where a large amount of energy is consumed.

Although old equipment has been replaced by more efficient models such as variable speed AC motors and drives [1], to earn more energy savings we propose intelligent energy control systems[2] which utilize the predictive information of mill process control and predicted material properties such as tensile strength, and yield stress for wider range of facilities.

One of our solutions is to modify rolling conditions to reduce energy usage within a proper range given by a material properties prediction system [3][4] which we have developed. This approach will minimize the energy consumption in the whole hot strip mill line while retaining product quality. Also predictive information from the model control system can be used to reduce more energy than is possible with conventional control methods. The basic idea, system configuration, and calculation results of the system will be shown in this paper.

INTRODUCTION

As people are increasing their concerns about the issues of energy and CO_2 emissions, how to reduce energy usage and CO_2 emission is more focused on. In the developed nations, large amounts of energy are consumed in sectors of industry, transportation, home etc. on the demand side. For example in Japan, energy consumption in the industrial and transportation sectors was 42.6% and 23.6% of total respectively in 2008 according to the Japanese Ministry of the Environment[5] as shown in Figure 1. Also the steel industry still consumes nearly 30% of energy in the industrial sector even though the ratio has gradually decreased. These data show the necessity of efforts by leaders of the steel industry to reduce energy usage in steel production.

As one of the electric suppliers of electric equipment and control systems to iron and steel makers, one of our missions is to contribute to saving energy in our customers' plants. In the steel industry large efforts to reduce energy consumption and CO_2 emissions have been made in the past. For example heat from blast furnaces is re-used for making electricity, and low efficiency machines have been replaced by high efficiency ones. Especially upstream processes such as blast furnaces, and arc furnaces use very large amounts of energy, so the room for saving energy is also large.

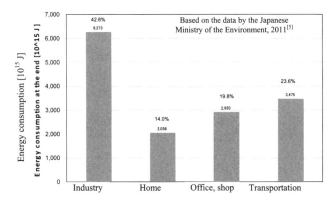

Figure 1. Energy consumption at demand end in Japan, 2008

Sometimes it is more important for the sake of producers' profits to retain product quality and production than to save energy in production lines. If the production lines cannot make target products with good quality, they make unsalable products while consuming large amounts of energy. Also it is important to protect equipment and facilities from any damage.

One of the solutions to save more energy without harming product quality and production is to consider the plant as a whole rather than just a single piece of equipment, and to utilize predictive information given by upper level control system. We call it "intelligent energy saving system".

Intelligent energy saving methods which retain product quality such as target dimensions and material properties in hot strip mills will be introduced and discussed in this paper.

Objective plant controlled by intelligent energy saving system

Figure 2 shows the configuration of a hot strip mill. Slabs are heated in a reheating furnace to about 1250°C. After descaling the slab surface, the slab is rolled several times in Roughing mills (hereinafter RM) and vertical mills to achieve the desired thickness and width. After crops at head/tail part are cut and the second descaling is applied, the bar, as it is now called instead of slab, goes into finishing mills (hereinafter FM) to get desired product dimensions i.e. thickness, profile of strip section and flatness. Also a target temperature at the FM exit is given and achieved by FM temperature control. After rolling in the FM the strip is cooled down to the target coiling temperature (hereinafter CT) on the run out table (hereinafter ROT) and coiled at the down coiler.

Table I shows the main control functions in the hot strip mill. In each function, setup functions set the initial status and dynamic control functions follow it to keep the target values.

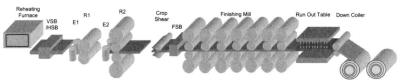

Figure 2. Configuration of hot strip mill

Table I. Main control function in hot strip mill

Control function	Purpose: control objective	Facility / manipulator
Slab temperature control	Target slab temperature	Reheating furnace
RM Setup (RSU)	Bar thickness and width	RM/roll gap, speed
Automatic Width Control (AWC)	Bar width	Vertical edger/roll gap
FM Setup (FSU)	Strip thickness	FM/roll gap, speed
Shape Setup (SSU)	Strip profile, flatness	FM/roll bender, roll shift
Automatic Gage Control (AGC)	Strip thickness	FM/roll gap
FM Delivery Temperature Control (FDTC)	FM delivery strip temperature	FM/speed, sprays
Coiling Temperature Control (CTC)	Coiler entry strip temperature	ROT/water

Figure 3 shows level of controls of rolling mill and energy saving, which is commonly used in steel industries. Level 3 makes production and management plan in business computers. Computer control and mill setup calculation are done in level 2 computers. Dynamic control functions are installed and executed in level 1 controllers at higher speeds than level 2. Usually a programmable logic controller, PLC, is used in level 1. Level 0 is the lowest level controllers such as motor drives, hydraulic controllers etc.

Functions shown in Table 1 are classified as level 2 and level 1; generally, level 2 functions are slab temperature control, RSU, FSU, SSU, FDTC and CTC [6], level 1 functions are AWC, and AGC.

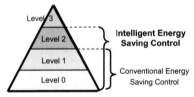

Figure 3. Level of rolling mill control and energy saving control

We define "intelligent energy saving control" as an energy saving control which uses predictive information given by level 2 functions or optimizes energy in across a broader range of facilities rather than just in a few pieces of equipment. Conventional methods use approaches like replacing old equipment with high efficiency models, replacing existing fixed speed motors and pumps with new variable speed ones. Conventional energy saving methods focus on the performance of individual pieces of equipment, so the energy savings can be realized in the Level 0 controller functions. But the intelligent methods use level 2 computer information, e.g. how much water will be used for the coming strip or how many rolling passes will be applied to the slab. The intelligent energy saving control is considered in the next chapter.

Energy saving system for whole hot strip mill line

There are two kinds of energy which are used in hot strip mills, fossil fuels in reheating furnaces and electricity in motors, drives, and controllers. In the furnace, slabs are heated up to the target temperature by burning heavy oil or natural gas. Slabs are rolled to their target thickness and width by electric motors in RM and FM. Also they are cooled down by water whose potential energy is given by pumps and motors.

Generally the ratio of fuels and electricity used in hot strip mills is about 6 to 4 without considering electric power generation efficiency. General efficiency of electric power generation is 50% or less, so the above ratio turns to 6 to 8 if the efficiency is considered.

Steel makers often want to reduce both fuel and electricity consumption at the same time. They also sometimes want to decide the priority of reducing either fuel or electricity according to management and business measures or operational constraints. For example,

(1) The management of steel makers just wants to reduce costs of fuel and electricity.

(2) The government may request them to reduce CO_2 emissions and for political reasons or for reasons of social responsibility, the management must comply with the request.

(3) By contract, rolling mills may not use more electricity than a pre-decided amount because of limited power generation or distribution capacity in the area around the steel plant. Penalties must be paid in case of exceeding the limitation and the management wants to avoid paying extra money.

Thus requests for energy savings are varied by the situation where the steel maker is located. Based on customer requests, both total energy reduction and the balance of energy used should be considered in hot strip mill.

In any of the above cases, it should be recognized that product quality and production may sometimes have priority over saving energy. Thus our approach is,

(a) Building a prediction system to estimate how much of energy saving effect variable factors will have - Energy Consumption Prediction System (hereinafter ECPS).

(b) Using Material Property Prediction System (hereinafter MPPS)[3][4] to check that material properties such as yield stress (hereinafter YS) and tensile strength (hereinafter TS) are retained.

(c) Making a concept for building an Energy Consumption Optimization System in hot strip mills to combine ECPS and MPPS.

ECPS (Energy Consumption Prediction System)

Visualization of energy consumption in the hot strip mill is very useful for understanding how much energy per coil is used. ECPS is one of the model functions such as RSU and FSU, and it predicts energy usage in each facility for the next rolled coil using initial setup calculation results by RSU, FSU etc. ECPS can be used as a simulation tool to change rolling conditions such as transfer bar thickness and target FM delivery temperature (FDT). For example if the operator is allowed to change target FDT of a coil from 890°C to 870°C, it will save energy for the coil. He can check how much energy will be saved by ECPS in advance of actual rolling.

Figure 4 shows an HMI example of ECPS calculations. Basic calculations are done according to Primary Data Input (PDI) which is given by the level 3 computer. But some rolling conditions can be modified and the results are displayed in "Trial ECPS Result" shown in right-bottom part of Figure 4. When a value is input to this part, ECPS and model setups calculate again interactively.

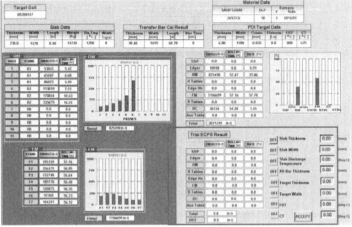

Figure 4. HMI example of ECPS

ECPS uses the prediction results of rolling torque and speed by RSU and FSU. Figure 5 shows an example of the predicted speed pattern in the FM before actual rolling starts. The speed is divided in several segments such as S4, S5. After NO. 7 stand (F7) is loaded on, the first acceleration is activated in S6 and S7, the second acceleration is done in S9. Before F7 off, deceleration starts and finally the tail of strip goes into the coiler at the end of S15.

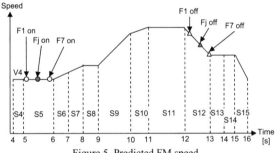

Figure 5. Predicted FM speed

Based on the speed pattern shown in Figure 5, energy consumption of the FM main motors is calculated as follows;
- Energy consumption E_{FM} at FM stand No. j [kJ]

$$E_{FM}[j] = Z_{EC}[j] \int_0^T P_W[j](t)dt = \sum_{i=js}^{je-1} \frac{(P_W[j](i) + P_W[j](i+1)) \cdot t_{seg}[j](i \sim i+1)}{2} \quad (1)$$

- Rolling power P_W [kW] at segment NO. i in Figure 5.

$$P_W[j](i) = \frac{1000 \cdot G_R[j](i) \cdot V_R[j](i)}{R[j]} \tag{2}$$

where

t_{seg} $(i \sim i+1)$: Transport time from segment IO. i to i+1 in Figure 5 [s],
G_R : Roll torque including loss torque [kIm],
V_R : Roll peripheral speed [m/s],
R : Roll radius [mm],
Z_{EC} : Adaptation term prepared as a future function [-],
js, je : Start/End segment number of stand j.

Table II shows calculation example which FDT or transfer bar thickness is changed.

Table II. Simulation results of ECPS

	Base case	Target FDT-20°C	Target FDT+20°C	Bar thickness -3mm	Bar thickness +3mm
Energy in mills [MJ]	2077.2	2090.7 (+0.65%)	2067.2 (-0.48%)	2093.9 (+0.80%)	2074.6 (-0.13%)

(*) Condition of base case: Slab thickness/width=230/1470mm, length=8.4m, Bar thickness=30.48mm,
Coil thickness/width=6.0/1390mm, Target FDT=880°C. Iote that value in ( ) is ratio to the base case.

In other sections of the mill, main motors of the slab sizing press, RM, and edgers, down coilers, and the auxiliary motors of the transfer tables and ROT are considered as sources of energy consumption in hot strip mills by adequate model equations. Currently some functions such as energy calculation in furnace are not considered, but those will be installed shortly.

MPPS (Material Property Prediction System)

In addition to saving energy, the production line is required to make good products according to given customer specifications and to avoid waste. The rolling models and ECPS and its trial modification function can calculate reasonable results from the view point of dimensional specifications. But they do not calculate the effects of changing temperature targets, strip speed and thickness reduction in each stand, i.e. strain rate and strain, which are strongly related to material properties. Thus it is necessary to check the results of ECPS from the material properties point of view. This is done using MPPS.

Originally the MPPS itself has been developed as an online calculation tool to monitor and check material properties of the products by using actual rolling data. MPPS has been applied to some hot strip mills and it works well. Figure 6 shows the MPPS configuration. The level 2 computer calculates setup data to produce the desired coil according to PDI given by the level 3 computer. After rolling actual data are acquired by the level 2 computer and they are given to the MPPS computer to predict material properties using the MPPS models. The predicted results are shown on HMI screen. The quality control unit in the hot strip mill measures actual material properties for sampled coils. The measured data are used for adapting MPPS models.

Examples of positive effects of MPPS:

(A) Frequency of tensile tests can be reduced because MPPS shows all coil material properties.
(B) Rolling and cooling process can be improved by using MPPS as an offline simulator.
(C) Production process can be optimized for smaller variation of material properties by MPPS.

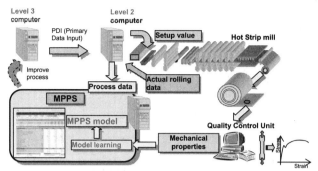

Figure 6. MPPS configuration

Figure 7 shows the MPPS model configuration. The reheating furnace model calculates γ-diameter by the initial structure model which describes grain growth during slab heat up in the furnace. The hot deformation model contains the expressions of recovery, re-crystallization, and grain growth after deformation. It uses γ-diameter from the reheating furnace model and actual rolling data in the RM and FM to calculate γ-diameter and dislocation density. The transformation model calculates micro-structure information such as α-diameter according to measured temperature and cooling rate on the ROT. Finally the structure-mechanical properties relationship model predicts mechanical properties such as YS and TS.

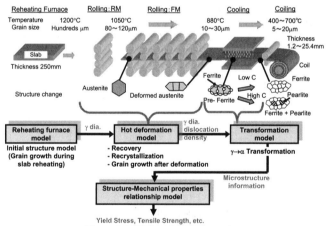

Figure 7. MPPS model configuration

MPPS can be used as both an online prediction tool and offline simulation tool. As an offline simulation tool, for example, it is possible to check the effect of temperature and cooling rate on YS. One of the results is shown in Figure 8.

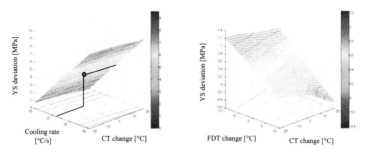

CT, Cooling rate and YS (FDT=860°C)   CT, FDT and YS (cooling rate=30°C/s)
Calculation condition: Target FDT=860°C, Target CT=560°C, Strip thickness=2mm, Carbon%=0.14wt%

Figure 8. Example of temperature effect on YS by MPPS as offline simulator

MPPS is used to check the results of ECPS as an offline simulation tool. Rolling models and ECPS pass the calculation results to MPPS to calculate material properties. Figure 9 shows an example of MPPS calculation results such as YS, TS and graphs of grain size, volume fraction, etc. An engineer can review them and check if they are reasonable or within the targets. If not, he can modify model inputs and ECPS conditions.

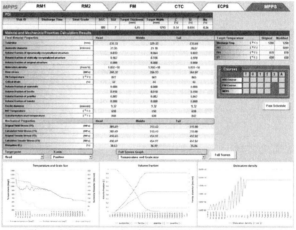

Figure 9. HMI example of MPPS to check ECPS results

Optimization

Our next step is to develop online automatic modification to find the optimal path to minimum energy while retaining quality. As already shown in Figure 4 and Figure 9, setups, ECPS and MPPS can be used to seek the optimal conditions as an offline simulation tool in actual operation to save energy and to retain product quality. Figure 10 shows the configuration of automatic and online optimization of rolling conditions. The automatic optimization function uses the results of ECPS and MPPS to get optimal condition for both energy and quality.

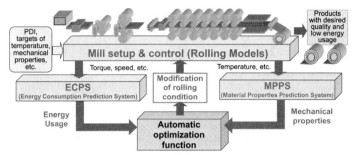

Figure 10. Configuration of optimization of rolling condition

Energy saving system for auxiliary motors

As defined above the "intelligent energy saving system" uses predictive information given by the upper level control system. One of the applications to auxiliary motors is for the ROT feed pump energy saving system.

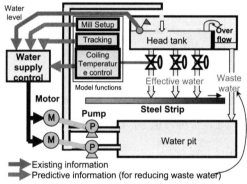

Figure 11. Configuration of ROT pump intelligent energy saving and CTC

The basic idea of ROT pump energy saving is to reduce surplus water to be fed into a head tank using predictive information from CTC in level 2. Figure 11 shows the configuration of ROT pump intelligent energy saving and CTC. Water to be used to cool down strips is pumped up from a scale pit to a head tank. Water volume in the scale pit is considered large enough to supply water. Water volume in the head tank should be kept at more than a certain level to provide an adequate water pressure. To do this it is necessary to overflow water in the head tank. But the overflow water is a waste and consumes excess energy of feed pumps.

To reduce overflow, necessary water to cool down strips should be pumped up at the necessary timing. Level 2 functions including CTC know how much and when water will be used for the strip. So if such predictive information is used adequately, the overflow water can be reduced.

The principal of energy saving for pumps is that power at the pump shaft is proportional to the cube of pump revolution.

$$Q \propto N \qquad (3)$$

$$P_W \propto N^3 \qquad (4)$$

where

Q :   Water flow pumped up [m³/s],
P_W: Power at pump shaft [kW],
N :   Revolution of pump [1/s].

These equations imply that slow change will reduce power and energy. In case a pump is operated at 50% speed, saved energy compared with 100% operation is derived as follows:

(1) Water flow is 0.5 times, and time of pumping up is 2 times 100% operation.
(2) Shaft power is $(0.5)^3 = 0.125$ times 100% operation.
(3) Total energy is "Power × Time", so it is $0.125 \times 2 = 0.25$ times 100% operation.

Assume a certain volume of water will be needed at time 't0' and case (A) needs 100% energy as shown in Figure 12. Case (B) only needs 25% of (A) based on the above simple calculation.

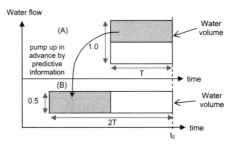

Figure 12. Basic idea of saving energy with predictive information

A simulation study is done to check the effect of intelligent energy saving of ROT pumps.
- Simulation condition
Pump: flow 2170[m³/h], head 20[m], 185[kW], 4 sets of pump and motor are available.
Head tank: capacity 132[m³], water volume is allowed to reduce by 106[m³], 80%.
Effective water head: 17[m].
Used water amount by CTC: shown in Figure 13.
Simulation time: 3600[s], i.e. one hour.
- Case setting
Case 1: Intelligent control with one fixed speed pump and three variable speed pumps.
Case 2: Conventional operation with four fixed speed pumps.
Case 3: Conventional operation with three fixed speed pumps (no graph in Figure 13).
- Result
Case 1: 250[kWh], Case 2: 880[kWh], Case 3: 660[kWh].
Energy usage of Case 1 is 29% of Case 2, 38% of Case 3,
i.e. energy saving of Case 1 against Case 2 is 71%, against Case 3 is 62%.

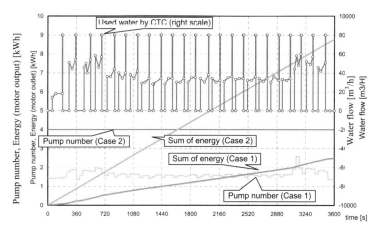

Note: Pump number is an output power of inverter drives.(e.g.) Pump number 1.5 = 3 pumps with 50% output.

Figure 13. Simulation result of ROT feed pump by intelligent energy saving control

Case 1 is much better than the others. Also if there is longer idle time between rolling, or water flow used by CTC is smaller, then intelligent energy saving control will have better effect. This method can be applied to not only ROT feed pumps but also descaling pumps etc.

CONCLUSION

Energy saving control which utilizes knowledge of the objective plants—or predictive information is effective to earn greater energy savings. One of the control methods is to combine rolling models, ECPS, and MPPS interactively in hot strip mills. Using this method better rolling conditions can be found which both save energy and retain product quality. The concept of automatic online modification is shown as our next goal. One application in hot strip mills is an intelligent energy saving control for pumps and motors. The result of the simulation study for ROT feed pumps shows the benefits of this method. We will develop more efficient and powerful energy saving control from the view point of optimization at a higher level.

REFERENCES

[1] Y. Sakiyama, and T. Takahasi, Introduction of Energy Saving Control System for Metal Industries, *Taiwan 2008 International Steel Technology Symposium*, C25, p 1-11 (2008)

[2] Y. Sakiyama, Advanced Energy Saving Technologies for Steel Industries, *Seminar on Energy Conservation and Environment Protection Technology in Steel Industry* (2010)

[3] K. Ohara, M. Chen, Y. Pan, M. Tsugeno, K. Honda, and M. Kihara, System for Predicting the Material Properties of Hot Rolled Steel, *Taiwan 2008 International Steel Technology Symposium*, A15, p 1-10 (2008)

[4] K. Ohara, M. Tsugeno, T. Tezuka, M. Sano, and J. Yanagimoto, A prediction and control system for the material properties of hot rolled steel, *AISE Annual convention* (2002)

[5] Environmental Statistics, *The Japanese Ministry of the Environment web site, http://www.env.go.jp/doc/toukei/index.html,* as of June 2011

[6] H. Imanari, W. Deng, and J. Shimoda, Advanced Coiling Temperature Control System in Hot Strip Mills, *10th International Conference on Steel Rolling*, p 473-479 (2010)

HOT GAS CLEANING WITH GAS-SOLID REACTIONS AND RELATED MATERIALS FOR ADVANCED CLEAN POWER GENERATION FROM COAL

Hiromi Shirai, Hisao Makino
Central Research Institute of Electric Power Industry
Yokosuka, Kanagawa, Japan

ABSTRACT

It is important to develop the hot gas cleaning technology with gas-solid reactions, which will make it possible to achieve the high removal performance of pollutants and the both reduction of waste water and heat loss. In integrated coal gasification combined cycle power generation (IGCC) and integrated coal gasification fuel cell power generation (IGFC), the development of hot gas cleaning technology is expected to lead to higher thermal efficiency. We have developed desulfurization, halide removal and mercury removal technologies. In this paper, we introduce those technologies as well as the related materials.

1. INTRODUCTION

Coal is an important energy source since it is the most abundant of the fossil fuels. However, advanced cleaning technology, for flue treatment, waste water treatment, ash treatment and so on is necessary for coal to be used efficiently in a power generation. The development of an advanced gas cleaning technology with gas-solid reaction is expected because such a technology can not only remove pollutants efficiently but also reduce waste water and heat loss. In pulverized coal-fired power plants, desulfurization unit with activated carbon has been used to reduce the waste water.

On the other hand, the practical use of integrated coal gasification combined cycle power generation (IGCC) technology, which leads to higher efficiency than that of pulverized coal-fired power generation technology, has been advanced. In IGCC, the hot gas cleaning technology with gas-solid reaction, in which dust, sulfur compounds (H_2S, COS), halides (HF, HCl) and so on are removed at 400 – 500 °C has also been developed in Japan. This technology can reduce heat loss and waste water, and will also be applied to integrated coal gasification fuel cell (MCFC, SOFC) power generation (IGFC) technology. In the hot gas cleaning technology, it is important to prepare a sorbent composed of solid materials. At our institute, sorbents for desulfurization, halides removal and mercury removal technologies have been developed. In this paper, we introduce these technologies as well as the related materials.

2. COAL GASIFICATION POWER GENERATION SYSTEM

An IGCC power generation system is shown in Figure 1. In an IGCC power plant with an air-blown coal gasifier, pulverized, dried coal is fed to the gasifier with air, and gasified. Coal gas, which is a combustible gas containing H_2 and CO, is exhausted from the gasifier. The gas is fed to the gas cleaning system, in which the dust and sulfur compounds are removed. The clean coal gas is burned in a gas turbine to generate the electric power. The heat of the gas exhausted from the gas turbine changes water into steam in a heat recovery steam generator, and electric power is generated by a steam turbine. Furthermore, IGFC will be developed to get the highest efficiency in a coal-fired power generation.

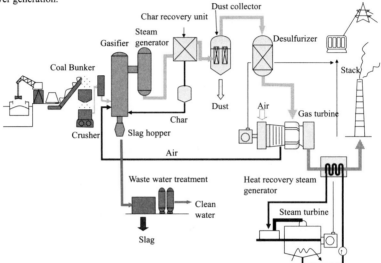

Figure 1 Integrated coal gasification combined cycle power generation system

3. GAS CLEANING SYSTEM

A wet scrubbing system has been introduced in commercial IGCC power plants, as shown in Figure 2. In this system, the dust is removed by a porous filter, which is made from alloy (Al-Fe, etc.) or ceramic (SiC, etc.). Then, the gas is cooled and water-soluble pollutants (HCl, HF, NH_3, etc.) are removed by a water scrubber. Sulfur compounds (H_2S, COS) are removed in an absorber with an organic solution (methyl di-ethanolamine (MDEA), etc.). However, considerable heat is lost in cooling the coal gas to below 40 °C. To get higher efficiency, the development of a hot gas cleaning system is required.   Our hot gas cleaning system is shown in Figure 3. The dust is removed by a porous filter at 400-500 °C. Then H_2S and COS are removed by hot gas desulfurization. In IGFC, it is necessary to remove the halides (HCl, HF), which deteriorate the fuel cell. This system not only has lower heat loss but also is simpler than the wet scrubbing system.

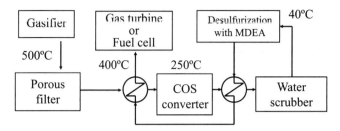

Figure 2 Conventional wet scrubber gas cleaning process (wet process)

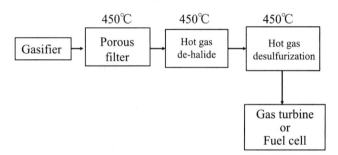

Figure 3 Hot gas cleaning process (dry process)

4. HOT GAS CLEANING TECHNOLOGY AND RELATED MATERIALS

4.1 Desulfurization

A metal oxide is used as the removal material. The reaction cycle is shown in Figure 4. The metal oxide is reduced to a stable metal oxide in the coal gas atmosphere, and the metal oxide absorbs the sulfur compounds, causing it to be converted into the metal sulfide. In the regeneration stage, the metal sulfide reacts with O_2, and is returned to the metal oxide. It is important to select a metal oxide with the following characteristics.

1) The metal oxide must be able to remove sulfur compounds so that its concentration reduced to below the target value. In IGCC, the target value is set on the basis of an environmental standard. In IGFC, the target value is set on the basis of the tolerance of the fuel cell. The target value in IGFC is less than 1 ppm.

2) The metal oxide must be regenerable. The sorbent containing the metal oxide must be reusable.

3) Carbon must not be deposited in the catalytic reaction of the metal. The deposited carbon destroys the sorbent containing the metal oxide.

Iron oxide (Fe_2O_3) and zinc ferrite ($ZnFe_2O_4$) were selected as removal materials on the basis of analysis of the chemical equilibrium and the results an experimental study.

(1) Iron oxide sorbent[1, 2]

Fe₂O₃ is reduced to Fe₃O₄ in the coal gas. Although the equilibrium removal concentration of Fe₃O₄ is affected by the gas composition, Fe₃O₄ can remove both H₂S and COS to concentrations of about 20 ppm in the coal gas formed in our air-blown gasifier. The reactions are follows.

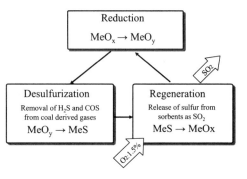

Figure 4 Reaction cycle

Reduction: $3Fe_2O_3 + H_2$ or $CO \rightarrow 2Fe_3O_4 + H_2O$ or CO_2

Desulfurization: $Fe_3O_4 + 3H_2S + H_2 \rightarrow 3FeS + 4H_2O$

Regeneration: $4FeS + 7O_2 \rightarrow 2Fe_2O_3 + 4SO_2$

The honeycomb shaped sorbent containing Fe₂O₃ developed for IGCC is shown in Figure 5. The sorbent is composed of Fe₂O₃ and support materials (TiO₂, etc.). The Fe₂O₃ content is 20 wt%. The preparation process is shown in Figure 6. Fe₂O₃-SiO₂ particles with a porous structure are prepared by a precipitation method. To get a fine primary particles of Fe₂O₃ (Figure 7) and prevent the sintering of Fe₂O₃ particle, SiO₂ is added. This also prevents the deposition of carbon, because SiO₂ prevents Fe₂O₃ from being reduced to metallic iron, which causes the deposition of carbon. The Fe₂O₃-SiO₂ particles are mixed with the support materials and a binder. The mixture is molded into the honeycomb shape. The porous sorbent is prepared by calcining at 800 °C.

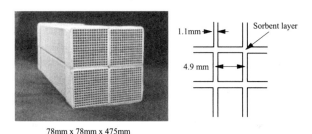

78mm x 78mm x 475mm

Figure 5 Honeycomb-shaped iron oxide sorbent

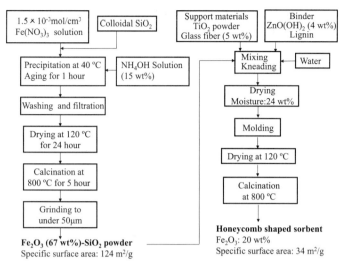

Figure 6 Preparation process for honeycomb-shaped iron oxide sorbent

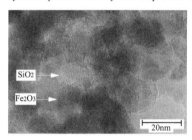

Figure 7 TEM photograph of grains in $Fe_2O_3\text{-}SiO_2$

(2) Zinc ferrite sorbent[3, 4]

Fe_2O_3 cannot remove the sulfur compounds to a concentration of below 1 ppm on the basis of chemical equilibrium. Although ZnO can reduce the sulfur concentration them below 1 ppm, it is difficult to regenerate it. Zinc ferrite is both removable and regenerable. The reactions are follows.

Desulfurization: $ZnFe_2O_4 + 3H_2S + H_2 \rightarrow 2FeS + ZnS + 4H_2O$

Regeneration: $2FeS + ZnS + 3O_2 \rightarrow ZnFe_2O_4 + 3SO_2$

A honeycomb-shaped sorbent with $ZnFe_2O_4$ was developed for IGCC and IGFC. The content of $ZnFe_2O_4$ is 20 wt%. The shape and preparation process are similar to those of Fe_2O_3.

(3) Removal performance

The removal performance of the honeycomb sorbents[5] was investigated using a fixed-bed reactor. The performance is shown in Figure 8. The normalization time is the ratio of the elapsed time to the breakthrough time, assuming that the sorbent absorbs H_2S instantaneously. This figure indicates that the iron oxide sorbent can remove the sulfur (H_2S+COS) to a concentration of below 100 ppm and that the zinc ferrite sorbent can remove the sulfur to a concentration of below 1 ppm. Furthermore, it was confirmed that these sorbents can be used repeatedly.

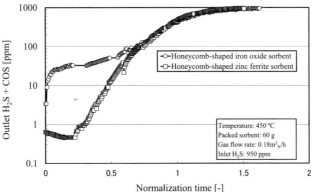

Figure 8 Characteristics of H_2S removal for each sorbent

(4) Fixed-bed desulfurization system[6]

The fixed-bed desulfurization system is shown in Figure 9. In each reactor, three steps, reduction, desulfurization and regeneration, occur sequentially in accordance with the time schedule. In the reduction step, a small amount of the coal gas not only reduces the sorbent but also decomposes the sulfates (Fe_2SO_4, $ZnSO_4$) formed in the regeneration step. In the desulfurization step, the sulfur compounds are removed from the coal gas. In the regeneration step, the O_2-containing gas is fed to the reactor. The sulfides contained in the sorbent react with O_2 and revert to oxides. In this system, the inlet O_2 concentration is set to 1.5% to maintain the maximum bed temperature below the calcination temperature of the sorbent since the regeneration is a strong exothermal reaction. This system has already been put into a practical use in a coal gasification pilot plant.

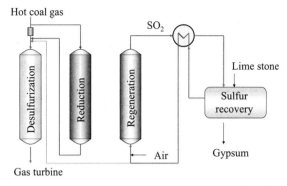

Figure 9 Fixed-bed desulfurization system

4.2 Removal of halides

In IGFC, the halides (HCl and HF) are removed before desulfurization to prevent the deterioration of the fuel cell. In the study of chemical equilibrium, it was found that sodium compounds (Na_2CO_3, $NaAlO_2$) remove halides to a concentration below 1 ppm at 400-500 °C. However, it is very difficult to regenerate the sorbent. Therefore, the sorbent is used as a disposable sorbent.

(1) $NaAlO_2$ sorbent[7, 8]

$NaAlO_2$ was selected as the removal material. The reactions are follows.

$$2NaAlO_2 + 2HCl = 2NaCl + Al_2O_3 + H_2O$$
$$5NaAlO_2 + 14HF = 2Na_5Al_3F_{14} + Al_2O_3 + 7H_2O$$

Cylindrical pellets of the sorbent for halide removal were developed for use in a fixed-bed reactor or a moving-bed reactor as shown in Figure 10. The sorbent is composed of $NaAlO_2$ and support materials (Al_2O_3, glass fiber etc.). The Na content is 21.8 wt%. The preparation process is shown in Figure 11. First, fine Al_2O_3 powder and glass fibers are added to Na_2CO_3 solution. The moisture contained in the mixture is reduced to a suitable concentration by evaporation to enable the molding of cylindrical pellets of the sorbent. The molded sorbent is dried and calcined at 700 °C.

Figure 10 Cylindrical $NaAlO_2$ sorbent

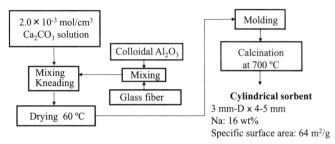

Figure 11 Preparation process for cylindrical NaAlO₂ sorbent

(2) Removal performance[7]

The removal performance of the NaAlO₂ sorbents was investigated using a fixed-bed reactor. The performance is shown in Figure 12. It is confirmed that the sorbent can remove both HCl and HF to concentrations below 1 ppm. The sorbent is not regenerable: therefore, it may be necessary that Al element is recovered from the sorbent then reused in the sorbent to reduce the cost of the sorbent.

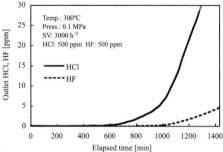

Figure 12 Characteristics of simultaneous removal of HCl and HF

4.3 Removal of mercury

If a flue gas emission standard for mercury is set for coal-fired power generation plants, mercury removal technology may be necessary. Mercury exists as a vapor of elemental mercury (Hg^0) in coal gas. The hot gas cleaning system with the mercury removal[9] is shown in Figure 13. The system is more complicated than that without the mercury removal because it is impossible to remove Hg^0 at 400-500 °C. The copper-based sorbent, which can remove Hg^0 at 120-160 °C, was developed at our institute.

(1) Copper-based sorbent[10, 11]

Cylindrical copper-based sorbent for use in a fixed-bed reactor or moving-bed reactor, were developed as shown in Figure 14. Copper element is in the form of CuO in the sorbent. The sorbent is sulfided as a pretreatment since copper sulfides such as CuS and Cu₂S absorb Hg^0. Although the absorption mechanism is not clear, it is considered that the absorbed mercury exists as HgS in the sorbent. The sorbent is regenerated by an oxygen-containing gas (O_2 2%) at 250 °C. The released mercury is captured by activated carbon at a temperature of below 60 °C.

The preparation process of copper-based sorbent is shown in Figure 15. First, a SiO_2-sol (Cu/Si=1.5) was added to $Cu(NO_3)_2$ solution. Then NaOH solution was added to the solution to precipitate the hydroxide. The hydroxide was then filtered, washed with distilled water, molded, air-dried and calcined in air at 300 °C.

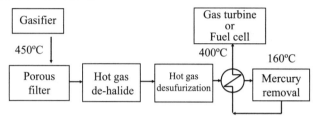

Figure 13 Hot gas cleaning process with mercury removal

Figure 14 Copper-based sorbent

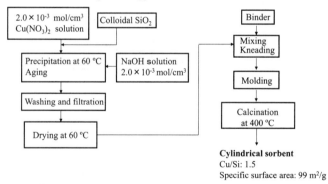

Figure 15 Preparation process for Copper-based sorbent

(2) Removal performance[10, 11]

The removal performance of the sorbents was investigated using the fixed bed reactor and is shown in Figure 16. The sulfided copper-based sorbent can remove Hg^0 to a concentration of below 0.1 $\mu g/m^3_N$, which is the detection limit of our mercury analyzer. The absorption capacity of mercury was greater than that of impregnated activated carbon. The removal performance after regeneration is shown in Figure 17. It was found that the time during which the sorbent kept the Hg^0 concentration below 5 $\mu g/m^3_N$ increases and becomes stable as the number of absorption- regeneration cycles increases.

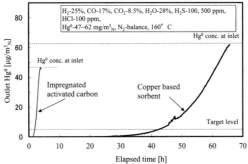

Figure 16 Characteristics of Hg^0 removal in each sorbent

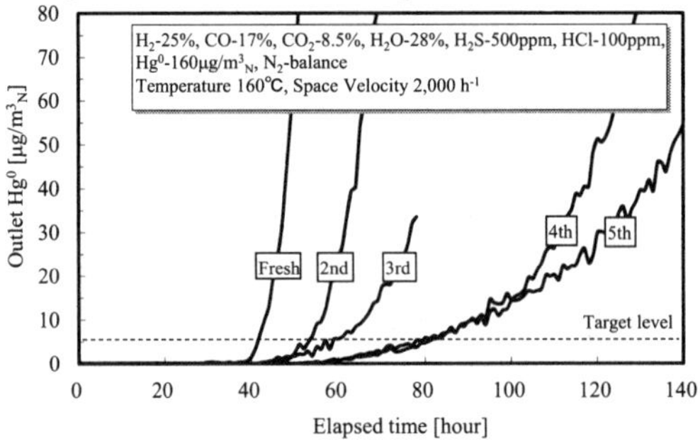

Figure 17 Change of characteristics of Hg0 by regeneration

4. CONCLUSION

To get higher efficiency in power generation using a gasifier, we have been developing hot gas cleaning technology with gas-solid reactions in which dust, sulfur compounds (H_2S, COS), halides (HF, HCl) and so on are removed at 400 – 500 °C. In the future, to practically apply this technology, it will be important to clarify the costs of construction and operation and the reliability of the technology.

REFERENCES

[1]H. Shirai, M. Kobayashi and M. Nunokawa, Reduction of Fe_2O_3-SiO_2 Particle in Air Blown Coal Gasification's Gas, Kagaku Kogaku Ronbunshu, 25, 714-720 (1999) in Japanese.

[2]H. Shirai, M. Kobayashi and M. Nunokawa, Modeling of Desulfurization Reaction for Fixed Bed system using Honeycomb Type Iron Oxide Sorbent and Desulfurization Characteristics in Coal Gas, Kagaku Kogaku Ronbunshu, 27, 771-778 (2001) in Japanese.

[3]H. Shirai, M. Kobayashi and M. Nunokawa, Characteristics of H_2S Removal of Mixed-oxide Sorbents Containing Fe and Zn at High Temperature, Journal of the Japan Institute of Energy, 77, 1100 - 1110 (1998).

[4]H. Shirai, M. Kobayashi and M. Nunokawa, Regeneration and Durability of Advanced Zinc Ferrite Sorbent for Hot Coal Gas Desulfurization, Proceedings of 5th International Conference of Gas Cleaning at High Temperature (2002).

[5]M. Nunokawa, M. Kobayashi and H. Shirai, Development of Regenerable Desulfurization Sorbent for Hot Coal Derived Gas, Proceedings of International Conference on the Characterization and Control of Interfaces for High Quality Advanced Materials (2003).

[6]H. Shirai, M. Nunokawa and T. Nakayama, Evaluation of Fixed-Bed Desulfurization System -Design of Demonstration Plant System and Estimate of Operating State, CRIEPI Report, W96004 (1997) in Japanese.

[7]M. Nunokawa, M. Kobayashi and H. Akiho, Development of Halide Removal Sorbent for Hot Gas Cleaning Technology, Proceedings of 4th International Conference on Clean Coal Technology for Our Future (2009).

[8]M. Nunokawa, M. Kobayashi, A. Yamaguchi, H. Akiho and S. Ito, Development of Gas Purification System for Multiple Impurities in Biomass/Refuse Derived Fuel Gasification Gases – Investigation of Preparation Method for Practical Molded Sodium Based Sorbent -, CRIEPI Report, M05017 (2005) in Japanese.

[9]M. Nunokawa, M. Kobayashi, Y. Nakao, H. Akiho and S. Ito, Development of Gas Cleaning System for Highly Efficient IGCC - Proposal for Scale-up Scheme of Optimum Gas Cleaning System Based on Generating Efficiency Analysis -, CRIEPI Report, M09016 (2010) in Japanese.

[10]H. Akiho, M. Kobayashi, M. Nunokawa, A. Yamaguchi, Y. Tochihara and S. Ito, Development of Dry Gas Cleaning System for Multiple Impurities -Proposal of Low-Cost Mercury Removal Process Using the Reusable Absorbent-, CRIEPI Report, M07017 (2008) in Japanese.

[11]H. Akiho, M. Kobayashi, M. Nunokawa, A. Yamaguchi, Y. Tochihara and S. Ito, Mercury removal from biomass-and waste-gasification gas with a reusable sorbent, Proceedings of Second international symposium on energy from biomass and waste (2008).

POLYALKYLENE CARBONATE POLYMERS - A SUSTAINABLE MATERIAL
ALTERNATIVE TO TRADITIONAL PETROCHEMICAL BASED PLASTICS

P. Ferraro
Empower Materials Inc.
New Castle, DE, USA

ABSTRACT

Polyalkylene Carbonates are revolutionary sustainable plastics made from carbon dioxide that consume considerably less petrochemicals than other conventional plastics. Petrochemical consumption is drastically reduced because carbon dioxide accounts for approximately up to 50% of the raw material feedstock. Therefore, the technology also potentially allows for sequestering greenhouse gases as the carbon dioxide source.

Empower Materials has developed a process utilizing a proprietary catalyst system to convert an epoxide and carbon dioxide into a polymer with many potential applications including coating and packaging for the food and beverage industries. Advancements have been made over the years to make the process more efficient and cost competitive. Catalyst development has also allowed for the production of a wider range of Polyalkylene Carbonates that includes Polyethylene Carbonate (QPAC®25), Polypropylene Carbonate (QPAC® 40), Polybutylene Carbonate (QPAC® 60), Polypropylene/cyclohexene Terpolymer Carbonate (QPAC® 100) and Polycylcohexene Carbonate (QPAC® 130). There has been work done to show that some of these plastics have excellent barrier film properties that make them suitable for packaging applications.

These polymers offer a green alternative to traditional polymers because of their reduction of petrochemical usage and potential reduction of greenhouse gases by its consumption in its synthesis.

INTRODUCTION

Copolymerization of carbon dioxide and alkylene oxide to high molecular weight Polyalkylene Carbonates was first synthesized by University of Tokyo 40 years ago. During the subsequent years, work has progressed in many areas including catalyst development to increase the conversion rate of the monomer to copolymer. These polymers exhibit unique chaining of the monomers by the alternate incorporation of carbon dioxide and epoxide moieties in the polymer chain. The desired reaction is the solvent based polymerization of an epoxide (propylene oxide or ethylene oxide) with carbon dioxide in the presence of a catalyst. There are a number of types of catalyst that have been investigated. It has been found that zinc based catalyst comprised of various types of metal carboxylates and acetates are very effective as catalysts. Zinc derivatives are believed to exhibit higher activity and lead to higher molecular weight polymers than

89

derivatives of cobalt or cadmium. Empower Materials has optimized the catalyst to extend the capability to produce very high molecular weight polymers efficiently. The general synthesis route for polyalkylene carbonates is as follows:

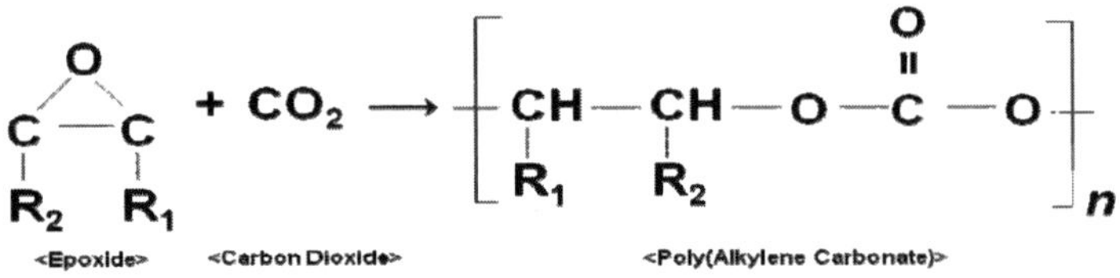

Figure 1- General Synthesis

The number of Polyalkylene Carbonate polymers that can potentially be made is large, a dozen or so varieties. Past and current development has focused on both Polypropylene Carbonate (PPC) and Polyethylene Carbonate (PEC).

There are several areas of use for these polymers. Since the high molecular weight polymers decompose cleanly at a low decomposition temperature, they find use in sacrificial binder applications for ceramic, glass or metallic particles in sintered molding applications. The Polyalkylene Carbonate is especially useful as a binder in temperature or contamination sensitive applications such as electronics where carbon or other residues are harmful.

A second application category that holds greater opportunities for large scale penetration is considered non sacrificial. The polymers are optically clear and mechanically stable for different applications. These polymers have strong barrier properties making them suitable in multilayer containers and films. In these applications, the polymer can be fabricated into films and other shaped articles and used in either multilayer's or blends with other polymers for various applications such as barrier film for packaging. These types of applications will be looked at more closely in this paper.

The overall objective for these polymers is to make a green polymer that has equal or better properties than traditionally petrochemical based polymer at a competitive cost.

RESULTS

The two primary Polyalkylene Carbonates that are commercially available as produced by Empower Materials and have been evaluated the most extensively are Polypropylene Carbonate, QPAC ®40, and Polyethylene Carbonate QPAC®25. Some common properties of both are summarized below in Table 1:

Table 1 – Polyalkylene carbonate properties

TYPICAL PROPERTIES	ASTM TEST METHOD	UNITS	QPAC® 40 PPC	QPAC® 25 PEC
Weight Average Molecular Weight	GPC		50,000-400,000	50,000-250,000
Density	D-792	g/cm^3	1.26	1.42
Shore Hardness	D-2240	D scale	79	NA
Water Absorption @ 23°C	D-570	%	0.4	0.4-0.6
Melt Flow Index @150°C/ 2,160g	D-1238	g/10 min.	0.9	1.4
Tensile Strength @ yield	D-638	psi	4500	670
Tensile Strength @ break	D-638	psi	1760	1700
Tensile Modulus	D-638	kpsi	300	NA
Elongation @ yield	D-882	%	3.5	NA
Elongation @ break	D-882	%	150	918
Izod Impact	D-256	ft-lbs/in	0.33	NA
Gardner Impact		In-lb	1.0	
T_g	DSC		35-40	10-28
Decomposition Temperature (onset)	TGA	°C	250	220
Heat of Combustion		Cal/gm°C	4,266	NA
Refractive Index	D-442		1.463	1.470
Haze		%	3.6	NA

PACKAGING APPLICATIONS

The critical properties in packaging are tensile and impact strength, optical clarity, heat sealability, chemical resistance and barrier properties. Typically, multiple films with 5-7 or more layers are used to improve flexibility while retaining barrier and clarity properties. The vinylidene chloride and vinyl chloride copolymers (formerly known as Saran) are difficult to fabricate but have outstanding barrier properties. The trade name, Saran, however is no longer composed of PVDC due to purported "environmental concerns with halogenated materials". It is now composed of polyethylene. Since they made this change, the material does not offer the same excellent barrier properties. Another commonly used material is Ethylene vinyl alcohol, EVOH, also known as EVAL. This plastic resin is commonly co extruded with other plastics and used in food applications. Another food packaging materials that is commonly used is PET. This is especially popular in bottle production. The comparison of PPC with PEC with some of these other polymers is shown below in Table 2:

Table 2 – Packaging Properties of PPC and PEC versus other polymers.*

Property	LDPE	Vinylidene Chloride	Polyester PET	PPC	PEC
WVTR– g/100sqin/24 hr/mil	1.0	0.2-0.6	1.7	5-15	5-15
Permeabilility to gases, cc/100 in²/atm CO₂ O₂ I₂	2700 500 180	4-44 0.8-7 .12-1.5	15-25 6-8 0.7-1	112 35 --	23 5 12
Resistance to grease and oils	Varies	Good	Good	Good	Excellent
Ease of Processing	Good	Poor	Good	Good	Good
Flexibility	Good	Good	Good	Good	Good
Tensile Strength, psi	1000-3500	8000-20000	30000	2000-5000	800
% Elongation	225-600	40-100	120-140	150	1000
Yield – sqin/lb/.001-inch	30000	16200	20000	20000	20000

*EVAL is not shown but has an O_2 barrier property similar or lower than Saran.

As the data shows, PEC is a significantly better O_2 barrier polymer than PPC. In comparison to other major film barrier candidates, PEC is at the upper end of the O_2 permeability range. As with many of the films utilized in the barrier film market, PEC would have to be an interior layer in multi layer film because it deforms easily with temperature due to its low Tg.

For water/ moisture barrier applications, both PEC and PPC are considered possible choices. Certain dry foods do not require maximum barrier properties. The properties listed in table 2 suggest that a single film of PEC and PPC might find a market niche in food packaging where cost and barrier properties are balance versus need, particularly for non perishable, dry food such as cereal box liners or potato chip bags. PPC and PEC has advantages over the other materials in several ways. They are clear and flexible and may have better interplay adhesion to other plastics and do not delaminate easily.

There has been previous work that has shown the gas barrier properties of Polyethylene Carbonate laminates. When combined with PE or other inferior barrier film materials, the PEC can greatly enhance the barrier properties:

Table 3 – Gas Transmission of PEC laminated with PE

Permeability to Gases, cc/100 in²/atm	PE/PEC	PE/PE	PE/PEC/PE
CO_2	10	---	13
O_2	1-3	60-70	1-2

Work continues in this area to evaluate multi layer systems using PPC or PEC as one of the layers. In many applications, the PPC or PEC is being combined with other renewable polymers that lack the barrier properties that these polymers offer.

The PPC and PEC can also be blended with other polymers. This data will not be shown in this paper.

DECOMPOSITION AND DEGRADABILITY

One of the great advantages of Polyalkylene carbonate copolymers is the way in which they burn or decompose. Typically, they burn in air with an almost invisible flame. The products that result from burning are CO_2 and water, both nontoxic substances present in the atmosphere. Other plastics, by contrast, produce cyanide, hydrocarbons or other toxic substances

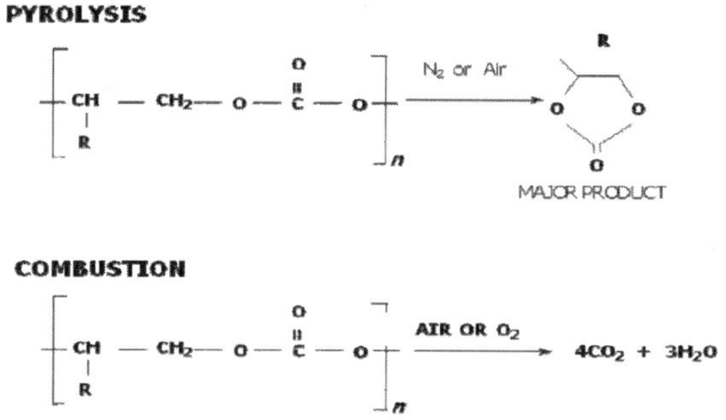

Figure 2- Decomposition Mechanism

Past degradation tests showed that degradation can be achieved in ambient conditions by blending acids or based into the polymer. The rate of decomposition is a function of the additive and its concentration in the sample. Other degradation tests indicate a positive weight loss of PEC in soil burial after 3 months. (Tests were not conducted on PPC)

Table 4 – Decomposition of PEC

Conditions	Results
PH 7.4 BUFFER FOR 40 DAYS[1]	No weight loss
Soil Burial (3 months) [2]	Positive weight loss
Clear-zone Test (30°C/15 days) [3]	Positive

CONCLUSIONS

Polyalkylene Carbonates are a welcome alternatives to traditional fossil fuel based plastics, since in addition to helping reduce the use of fossil fuels, they can create a market for waste carbon dioxide. They are a viable, eco-friendly alternative solution to carbon capture and carbon sequestration.

The Polyalkylene Carbonates have several positive features that make them useful in many commercial film applications. Although one of the main challenges would be in freestanding bottle or film applications because of its low heat distortion temperature, there are many combined laminate applications where the Polyalkylene Carbonates would be very effective. The material has a good melt flow index resulting in low processing cost. The O_2 barrier properties of Polyethylene Carbonate are comparable to other traditional film materials. They have good tear strength and good elongation. Polyalkylene Carbonates heat seals readily. Additionally the laminates possess enhanced gas barrier properties. Films that comprise several co-extruded polymeric layers that contribute individually to improving seal strength, impact strength, resilience, and reduced-temperature salability of the film can collectively improve the overall film's properties.

It is also a potential biodegradable plastic with potential ease of disposal via incineration since water and CO_2 would be the only products of complete combustion: no HCl, $COCL_2$, nitrogen oxides, etc. would be formed.

Currently, these polymers can be manufactured on a commercial scale by Empower Materials. However, additional scale up work is needed for economies of scale and further application testing is necessary. Pending success of this next stage, the Polyalkylene Carbonate family has great potential to displace other traditional petroleum based polymers, offerings substantial reduction in non renewable resources and the opportunity to capture CO_2 greenhouse gases.

REFERENCES

1.T. Kawaguchi, M. Nakano, K. Juni, S. Inoue and Y. Yoshida, Chem. Pharm. Bull. 31 (4), 1400-1403 (1983).

2 . H. Nishida, Y. Tokiwa, Chem Letter 1994 p 421

3. S. Yang, X. Fang, L. Chen, Polym. Adv. Series 1996, 7, 605

Materials for Nuclear Waste Disposal and Environmental Cleanup

CHARACTERIZING THE DEFECT POPULATION INTRODUCED BY RADIATION DAMAGE

Paul S. Follansbee
Saint Vincent College
Latrobe, PA, USA

ABSTRACT

Radiation damage in iron-based alloys can lead to an increased strength, accompanied by a decreasing resistance to fracture. While the vast majority of stress-strain curves measured on irradiated material have been at room temperature, quasi-static loading conditions, a few measurements have been performed at other loading temperatures and strain rates. These latter conditions enable an evaluation of the influence of the newly created defects on the deformation kinetics. A review of available measurements and an analysis of information that can be drawn from these measurements will be presented. A motivation of this work is to encourage more measurements of temperature and strain-rate dependent yield in irradiated materials.

INTRODUCTION

Materials exposed to neutron and / or proton irradiation undergo radiation damage. High energy particles create a cascade of microstructural damage through creation of defects. The coalescence of defects into defect structures and the effect of these structures on physical and mechanical properties is an area of considerable research[1,2].

Two comprehensive papers by Byun and Farrell have summarized radiation hardening and the ensuing plastic stability processes in polycrystalline metals[3,4]. The increase in yield strength measured in tensile specimens machined from irradiated material was observed to follow a power law relationship with distinctive low-dose and high-dose regimes of behavior[3]. The dose-dependence of the strain-hardening behavior in radiation-damaged metals suggested that radiation damage and typical dislocation-hardening yield similar microstructural effects[4].

The purpose of this paper is to review these results and measurements reported by other investigators and to analyze the behavior in context of a particular state-variable modeling approach that has found utility as a constitutive model in a variety of metals. Of particular interest are the deformation kinetics and the evolution of the irradiation-induced defect population. Ultimately, the goal is to propose a deformation model that takes account of irradiation damage.

HARDENING BY VACANCIES

Irradiation can lead to creation of individual vacancies, which coalesce into defect clusters. It is instructive to first analyze the effect that a population of vacancies has on strength. Meshii and Kauffman[5] measured yield strength as a function of test temperature on 99.999% pure cold wires quenched from different temperatures and at different quench rates. Because the equilibrium concentration of vacancies is strongly dependent on temperature, a rapid quench traps the vacancies – although these are expected to be collapsed into vacancy clusters or small voids. Figure 1 shows measurements for wires quenched from 1030°C at 30,000 °C/s. The quenched and aged samples show a much higher yield stress. However, the temperature-dependence of the yield stress is quite similar to that for the slow-cooled samples up to ~500 K. As will be shown, these observations mirror the behavior in radiation-damaged metals.

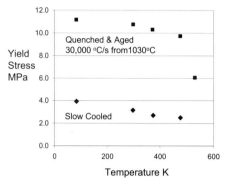

Figure 1. Yield stress measurements in slow cooled and quenched and aged gold (from Meshii and Kauffman[5]).

Analysis of Deformation Kinetics

It is assumed that the slow cooled samples and that the quenched and aged samples exist in two distinct "states". Measurement of the yield stress versus temperature can be used to probe the kinetics of the deformation process in each of these materials. For thermally activated deformation processes the following equation is introduced[6,7,8]

$$\hat{\sigma} = \sigma_a + \hat{\sigma}_\varepsilon + \hat{\sigma}_v \tag{1}$$

$$\frac{\sigma}{\mu} = \frac{\sigma_a}{\mu} + s_\varepsilon(\dot{\varepsilon},T)\frac{\hat{\sigma}_\varepsilon}{\mu_o} + s_v(\dot{\varepsilon},T)\frac{\hat{\sigma}_v}{\mu_o} \tag{2}$$

$$s_\varepsilon(\dot{\varepsilon},T) = \left\{1 - \left[\frac{kT}{\mu b^3 g_{o\varepsilon}}\ln\left(\frac{\dot{\varepsilon}_{o\varepsilon}}{\dot{\varepsilon}}\right)\right]^{1/q_\varepsilon}\right\}^{1/p_\varepsilon} \tag{3}$$

$$s_v(\dot{\varepsilon},T) = \left\{1 - \left[\frac{kT}{\mu b^3 g_{ov}}\ln\left(\frac{\dot{\varepsilon}_{o\varepsilon}}{\dot{\varepsilon}}\right)\right]^{1/q_v}\right\}^{1/p_v} \tag{4}$$

Equation 1 defines the state parameter – the mechanical threshold stress – as the sum of the threshold stress characterizing interactions of dislocations with each other and the threshold stress characterizing the interactions of dislocations with vacancies. An athermal stress σ_a is added to account for any dislocation – obstacle interactions that are not thermally activated (e.g., with grain boundaries). Equation 2 gives the yield stress at any temperature and strain rate as a function of the two state parameters. Equations 3 and 4 give the thermal-activation relation in terms of the Boltzmann constant k, the shear modulus μ, the shear modulus at 0 K μ_o, the Burger's vector b, the normalized activation energy g_o, a strain-rate constant, and the powers p and q. Clearly, there is insufficient data in Figure 1 to enable an independent evaluation of the parameters in equations 1-4. However, as an FCC metal, gold is similar to copper, which is the original material analyzed according to this model[6] (with well over 1000 measurements). Thus, several of the constants can be taken from the original work. In analyzing temperature and strain-rate dependent (when available) yield stress measurements, the standard procedure has been to create the plot shown in Figure 2. In this figure the model predictions are represented by the solid line. Model parameters are listed in Table I (along with those indicated in the labels in Figure 2).

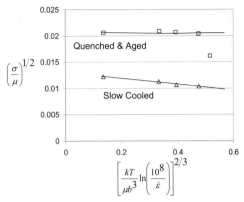

Figure 2. Meshii and Kauffman data analyzed according to Equations 1-4 with the model parameters shown in Table I.

Table I. Model parameters for predictions in Figure 2

Material	σ_a	$\hat{\sigma}_\varepsilon$	$g_{0\varepsilon}$	$\hat{\sigma}_v$
Slow cooled	0	4.6 MPa	3.5	0
Quenched & aged	7.3 MPa	4.6 MPa	3.5	0

Note that while Equation 1 includes the possibility of a threshold stress characterizing interaction of dislocations with vacancies - $\hat{\sigma}_v$, the best fit was obtained by not introducing this but instead by introducing an athermal stress of 7.3 MPa along with the same value of $\hat{\sigma}_\varepsilon$ as used in the slow-cooled case[*]. Acceptable agreement with the quenched and aged measurements could not be obtained simply by leaving the athermal stress as zero and increasing $\hat{\sigma}_\varepsilon$. Doing so led to a stronger temperature dependence than observed in the data. The conclusion is that the vacancy population introduces a temperature and strain-rate insensitive – athermal – obstacle. This conclusion is consistent with the notion that a void of more than a few Burgers vectors in size represents a strong obstacle with a large attractive force between a dislocation and the void.[9] It follows that typical thermal fluctuations would be ineffective in assisting the dislocation past the obstacle.

---

[*] For well annealed FCC metals, this threshold stress would be expected to start at zero and increase with strain hardening. The fact that a small value of 4.6 MPa was found to describe the data well may reflect the fact that the stresses in pure gold are an order of magnitude less than those observed in pure copper and nickel.

With the vacancy-case as background, radiation-induced damage in metals can be similarly analyzed.

MEASUREMENTS IN IRRADIATED METALS

While numerous room temperature, quasi-static (QS) yield stress measurements have been reported in irradiated metals, there are very few cases where a temperature and / or strain-rate dependent yield stress has been reported. The expectation is that irradiation damage – similar to vacancy hardening – will be relatively strain-rate and temperature insensitive; this is a result that should be independently verified. Fortunately, some measurements are available in the literature. Table II lists the data sources considered in this paper.

Table II. Data Sources Examined

Researchers	Material	Irradiation Condition	Dose	Tensile Test Conditions
Murty[10]	Mild Steel	Fast (> 1 MeV) neutrons	Unirradiated and 3.9×10^{16} to 1.4×10^{19} neutrons/cm^2	1.4×10^{-4} s^{-1} RT and 373 K
Albertini and Montagnani[11]	316L SS	Unspecified	Unirradiated and 2.2 dpa at 400°C	RT and 673 K 0.004 s^{-1} to 60 s^{-1}
Byun et al[12] (SNS 101060100)	EC316LN	800 MeV proton and spallation neutrons	Unirradiated and 0.54 dpa and 1.87 dpa; 58°C to 160°C	RT and 164°C 0.001 s^{-1}
Dai, Egeland, and Long[13]	EC316LN	SINQ targets	Unirradiated 3 dpa to 17.3 dpa at 357 K to 619 K	RT to 350°C (depending on dose temperature) 0.001 s^{-1}
Byun and Farrell[4]	EC316LN	800 MeV proton and spallation neutrons	Unirradiated 0.5 dpa to 10.7 dpa; 60°C to 160°C	RT 0.001 s^{-1}

The variety in irradiation conditions – particularly the irradiation temperature – introduces a potential complication. When considering the accumulation of damage, for example strain hardening or dislocation accumulation in copper or nickel, the temperature at which the damage is being introduced is important[6]. This will be examined for the case of irradiation damage below. While data on other materials is available from the references included in Table II, mild steel and 316 SS were chosen because these are materials for which extensive strain-rate and temperature dependent data – as well as strain-hardening data – is available and has been analyzed with the Mechanical Threshold Stress (MTS) model introduced above.

Deformation Kinetics in Mild Steel

As a BCC metal with a strong contribution to strengthening from a Peierl's barrier, Equations 1 and 2 must be modified. The modification includes additional terms (mechanical threshold stresses) representing the Peierls barrier and the interaction between dislocations and impurity or interstitial atoms (e.g., carbon)[14]:

$$\hat{\sigma} = \sigma_a + \hat{\sigma}_p + \hat{\sigma}_i + \hat{\sigma}_v \qquad (5)$$

$$\frac{\sigma}{\mu} = \frac{\sigma_a}{\mu} + s_p(\dot{\varepsilon},T)\frac{\hat{\sigma}_p}{\mu_o} + s_i(\dot{\varepsilon},T)\frac{\hat{\sigma}_i}{\mu_o} + s_v(\dot{\varepsilon},T)\frac{\hat{\sigma}_v}{\mu_o} \qquad (6)$$

In this case, the term representing interactions of dislocations with each other has been dropped since typical strain hardening is not of interest. Figure 3 shows, using a similar representation as illustrated in Figure 2, the model predictions along with the Murty data. Plotted along with the Murty measurements are measurements in 1018 steel from unpublished work at Los Alamos National Laboratory. The upturn at low values of the abscissa results from the contribution of the strong Peierls barrier, but at QS strain rates and RT and above, this obstacle loses effectiveness. Table 3 lists the parameters used for the model predictions shown in Figure 3.

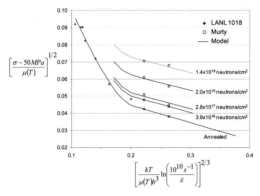

Figure 3. Yield stress in unirradiated and irradiated mild steel along with MTS model predictions.

Table III. Model parameters for predictions in Figure 3

Dose neutrons /cm^2	σ_a	$\hat{\sigma}_p$ $g_{op} = 0.10$	$\hat{\sigma}_i$ $g_{oi} = 0.6$	$\hat{\sigma}_v$ $g_{ov} = 0.6$
0	50 MPa	1930 MPa	300 MPa	0
3.4x10^{16}	50 MPa	1930 MPa	300 MPa	46 MPa
2.8x10^{17}	50 MPa	1930 MPa	300 MPa	64 MPa
2.0x10^{18}	50 MPa	1930 MPa	300 MPa	172 MPa
1.4x10^{19}	50 MPa	1930 MPa	300 MPa	300 MPa

Note that agreement between the model and the test results was achieved by varying a single parameter in the MTS model – $\hat{\sigma}_v$, which is the mechanical threshold stress characterizing interaction of dislocations with the irradiation-induced defect. Whereas the quenched and aged vacancy population in the Meshii and Kauffman gold wires appeared to produce an athermal obstacle, the strength of the irradiation-induced obstacle is weakly (as compared to the Peierls barrier) affected by temperature – as indicated by the g_{ov} value of 0.6. As expected, the strength of the irradiation-induced defect increases – evolves – with dose. Later, this dependence will be analyzed in more detail.

Deformation Kinetics in 316 Stainless Steel

Table II lists several sources of measurements in 316 stainless steels. One reason why this data was selected is that this author has recently completed an extensive assessment of the strength and variability of strength in the 304 / 304L / 316 / 316L systems[15]. This work was motivated by loading conditions at temperatures from RT to ~100°C and strain rates from QS to ~1000 s⁻¹. A "lower-bound" strength model was proposed for these conditions[15]. While the test temperatures in Table II are shown to be as high as 373°C , this model nonetheless gives a baseline model for analyzing the stress strain curves in 316 SS. It is worth noting that over the range of conditions investigated the variability between multiple lots of a single composition (e.g., 316L) exceeds the strength differences between different compositions (e.g., 304L versus 316L). Thus, while the different compositions provide benefits in high temperature (creep) strength or corrosion resistance, these alloys (and their mechanical properties) are essentially indistinguishable over the range of conditions of interest here.

Figure 4 shows measurements of the temperature and strain-rate dependent yield stress in 304 SS reported by Steichen[16] along with the model prediction. As an austenitic stainless steel, there is no Peierls barrier to include in Equations 5 and 6. Model parameters are listed in Table IV.

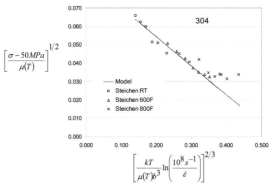

Figure 4. Yield stress versus temperature and strain rate in 304 SS reported by Steichen[16] along with the model prediction.

Table IV. Model parameters for predictions in Figure 4

Material Unirradiated	σ_a	$\hat{\sigma}_\varepsilon$ $g_{o\varepsilon} = 1.6$	$\hat{\sigma}_i$ $g_{oi} = 0.4$	$\hat{\sigma}_v$ $g_{ov} = 0.4$
304 SS	50 MPa	0 MPa	529 MPa	0 MPa
316 SS	50 MPa	0 MPa	715 MPa	0 MPa

The model agrees well with the data until values on the abscissa exceed ~0.35. At a QS strain rate, this corresponds to temperatures greater than ~600 K, where some investigators have reported the precipitation of $M_{23}C_6$ carbides. Since some of the measurements in Table II are at higher temperatures than this, caution is required in applying the model.

Included in Table IV are values for the three 316 stainless steels listed in Table II. A slightly higher value (than used for 304 SS) of the mechanical threshold stress characterizing interactions of dislocations with impurity / interstitial atoms led to a good fit with the data. Figure 5 shows the Dai measurements of yield stress on unirradiated and irradiated 316 and model predictions. Note that the higher temperature measurements at 7.6 dpa, 9.7 dpa, and 11.3 dpa are above the prediction – which follows from the discussion above. The prediction at 4 dpa and 5.3 dpa required $\hat{\sigma}_v$ values of 808 MPa and 1000 MPa, respectively. As was observed with the Murty data in Figure 3, agreement between the model and measurements required a "thermally activated" obstacle. In this case $g_{ov} = 0.4$.

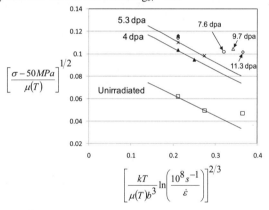

Figure 5. Yield stress versus temperature and strain rate in 316 SS reported by Dai et al[13] along with the model predictions.

INCORPORATING STRAIN HARDENING
The analysis to this point has been of the temperature dependence of the yield stress and comparison of predictions of the MTS model with measurements in mild steel and 316 stainless steel. All of the references in Table II, however, report full stress-strain curves. Byun and Farrell observed that the stress-strain curves on irradiated metals could be superimposed on the stress-

strain curve on the unirradiated metal simply by shifting the former curve along the strain axis, and that the samples with higher levels of damage were shifted farther along the axis. The implication of this is that the irradiation-induced hardening is analogous to strain hardening.

The MTS model describes the entire stress-strain curve, and it is useful to assess how the strain-hardening terms in the model can be modified to describe the stress-strain curve in irradiation-damaged metals. In the original MTS model, strain hardening is defined as the differential increase of the mechanical threshold stress characterizing interactions of dislocations with each other:

$$\frac{d\hat{\sigma}_{\varepsilon}}{d\varepsilon} = \hat{\theta}_o(\dot{\varepsilon})\left[1 - \frac{\hat{\sigma}_{\varepsilon}}{\hat{\sigma}_{\varepsilon s}(\dot{\varepsilon},T)}\right]^{\kappa} \tag{7}$$

$$\hat{\theta}_o(\dot{\varepsilon}) = A_o + A_1\dot{\varepsilon} \tag{8}$$

$$\ln\left(\frac{\hat{\sigma}_{\varepsilon s}}{\hat{\sigma}_{\varepsilon s0}}\right) = \frac{kT}{\mu b^3 g_{0\varepsilon s}}\ln\left(\frac{\dot{\varepsilon}}{\dot{\varepsilon}_{0\varepsilon s}}\right) \tag{9}$$

Without the power κ, Equation 7 is the Voce Law, where the threshold stress converges to a saturation value. Equation 9 describes the temperature and strain rate dependence of the saturation threshold stress.

Figure 6 shows how closely the predicted stress strain curves match the experimental curves of Dai et al at RT and 150°C but not at 350°C – for the reasons describe earlier. These predictions were made using the MTS parameters in Table IV (316 SS) along with the hardening parameters in Table V.

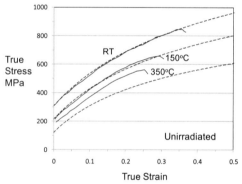

Figure 6. Stress-Strain curves in 316 SS reported by Dai et al[13] compared to the model predictions.

Strain Hardening in Irradiated Material

Following the approach summarized in Table IV and Figure 5, strain hardening in irradiated material is analyzed by increasing the $\hat{\sigma}_v$ value until the predicted curve matches the experimental curve with all other parameters (e.g., in Table V) held constant. Accordingly,

$\hat{\sigma}_v$ increases with increasing dose. The addition of a second evolving state variable introduces

Table V. Model parameters for predictions in Figure 6

316 SS	Parameter	Value
$\hat{\theta}_o$	A$_o$	3300 MPa
	A$_1$	20 MPa
κ	κ	4.4
$\hat{\sigma}_{ES}$	$\hat{\sigma}_{ESO}$	3250 MPa
	$\dot{\varepsilon}_{0ES}$	$10^7\,\text{s}^{-1}$
	g_{0ES}	0.25

complexity to the model. Byun and Farrell's conclusion that the irradiation-induced defect structure is similar to a strain-induced structure suggests a modification to the evolution equations.

$$\frac{d\hat{\sigma}_\varepsilon}{d\varepsilon} = \hat{\theta}_o(\dot{\varepsilon})\left[1 - \frac{\hat{\sigma}_\varepsilon + \hat{\sigma}_v}{\hat{\sigma}_{ES}(\dot{\varepsilon},T)}\right]^\kappa \tag{10}$$

$$\hat{\sigma}_v = f(dpa) \tag{11}$$

Figure 7 shows predictions of the model with the experimental results of Byun and Farrell. The dashed lines in this figure are the model predictions with $\hat{\sigma}_v$ values of 157 MPa, 193 MPa, 229 MPa, 268 MPa, and 358 MPa for doses of 0.5 dpa, 1.1 dpa, 2.5 dpa, 3.6 dpa, and 10.7 dpa respectively. It is evident in this figure that using the entire stress-strain curve rather than just the yield points is very instructive. The lower dose tensile curves, for instance, show a mild yield drop (although there is no real drop in stress) that would lead to overestimate of the irradiation-induced hardening if only the 0.002% yield stress were quoted.

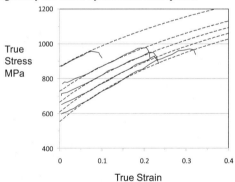

Figure 7. Stress-Strain curves in 316 SS reported by Byun and Farrell[4] compared to the model predictions (dashed lines). From lowest curves to highest, the doses were reported as 0.5 dpa, 1.1 dpa, 2.5 dpa, 3.6 dpa, and 10.7 dpa.

Similarly Figure 8 shows the Dai et al[13] and Byun et al[12] measurements (only the RT measurements are included) along with the model predictions. As in the Byun and Farrell curves, these were analyzed using the same methodology, with no parameter changes other than $\hat{\sigma}_v$.

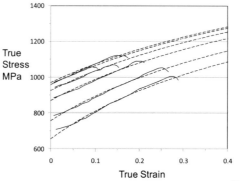

Figure 8. Stress-Strain curves in 316 SS reported by Dai et al[13] compared to the model predictions (dashed lines). From lowest curves to highest, the doses were reported as 3 dpa, 4 dpa, 5.3 dpa, 7.6 dpa, 9.7 dpa, and 11.3 dpa.

Table VI summarizes the results for variation of $\hat{\sigma}_v$ with dose for the data sets analyzed. Only estimates for tensile test temperatures less than 600 K are included. Cases where tensile tests at different temperatures on material of equivalent dose offer an estimate of the scatter in the estimates of $\hat{\sigma}_v$ since the values should be identical. In these cases the estimates agree to within ~15%. The Albertini and Montagnani estimates appear low compared to the others.

Table VI. Summary of dose and exposure temperature versus $\hat{\sigma}_v$ for the four data sets analyzed

Reference	Dose (dpa)	Exposure Temp (K)	Tensile Test Temp (K)	$\hat{\sigma}_v$ (MPa)
Byun and Farrell[4]	0.5	333 - 373	RT	157
	1.1			193
	2.5			229
	3.6			268
	10.7			358
Dai et al[13]	3	350 - 364	RT	222
	4	380 - 399	RT	286
	4		373	250
	5.3	415 - 440	RT	358
	5.3		423	336
	7.6	477 - 513	RT	393
	7.6		523	436
	9.7	538 - 584	RT	415

	11.3	582 - 637	RT	422
Byun et al[12]	0.54	333 - 373	RT	193
	0.54		437	236
	1.87	333 - 373	RT	329
	1.87		437	379
Albertini and	2.2	673	RT	93
Montagnani[11]	2.2		RT (27 s⁻¹)	93

Evolution Law for Irradiation Damage

Figure 9 plots from the results listed in Table VI $\hat{\sigma}_v$ versus dose. A point at zero dose and $\hat{\sigma}_v = 0$ is included. The curve is drawn according to a Voce law of the same form used for evolution of $\hat{\sigma}_\varepsilon$ in Equation 7:

$$\frac{d\hat{\sigma}_v}{d\,dpa} = \theta_{ov}\left(1 - \frac{\hat{\sigma}_v}{\hat{\sigma}_{vso}}\right)^{\kappa} \tag{12}$$

For the curve in Figure 9, $\theta_{ov} = 450\,MPa$, $\kappa = 5$, and $\hat{\sigma}_{vso} = 750\,MPa$. A fair amount of scatter is

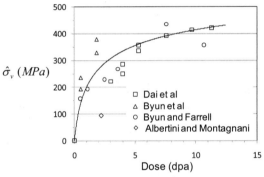

Figure 9. Evolution of $\hat{\sigma}_v$ with dose.

evident in Figure 9 and, perhaps, some systematic trends between data sets. Although Table VI indicates that the irradiation damage was imposed at various exposure temperatures, the data does not justify introduction of a temperature-dependent saturation stress, as is the case in Equation 9 for $\hat{\sigma}_{\varepsilon so}$. Based on these limited results, it is not possible to conclude exposure temperature affects damage evolution. It is interesting that the Albertini and Montagnani exposures were at 673K and yielded such a low value of $\hat{\sigma}_v$. While it is tempting to conclude the higher temperature led to the low threshold stress, other measurements at temperatures close to this are included in the figure (see Table IV).

Evolution in Mild Steel

Figure 10 plots $\hat{\sigma}_y$ from Table III versus dose. The curve again represents the Voce law with $\theta_{ov} = 3x10^{-16} MPa$, $\kappa = 5$, and $\hat{\sigma}_{vso} = 500\,MPa$. With only 4 data points, one may not have started with the Voce law, but it does match the data. The yield drops and Lüders band region in the curves reported by Murty[10] make it difficult to match the entire stress-strain curve as illustrated above for 316 SS.

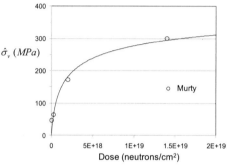

Figure 10. Evolution of $\hat{\sigma}_y$ with dose for mild steel.

DISCUSSION

The evidence that the irradiation-induced obstacle is a thermally activated obstacle is fairly strong. This follows from the model fit to the variation of yield stress with test temperature in Figures 3 and 5 as well as from the agreement between the full stress-strain curves and the model prediction shown in Figures 7 and 8. Assuming an athermal obstacle – as was found to be active in the coalesced vacancy population in quenched and aged gold – would not have enabled the agreement between model predictions and experimental results reported here. However, the conclusion is based on only a few test temperatures. It would be useful to include a wider variation of test temperatures (including below RT) in future experiments. Equally useful would be testing at other than QS strain rates.

Characterizing the entire stress-strain curve was useful. Doing so helped to explain the subtle difference in the rate of strain hardening observed when the yield stress of irradiated tensile test specimens was translated to the same stress level in an unirradiated tensile sample. The difference follows from the mild strain softening that follows yield. Interestingly, the comparisons in Figures 7 and 8 of this manuscript show good agreement between measurements and the predicted strain-hardening rates in irradiated material. Accurate prediction of the rate of strain hardening is essential for accurate prediction of strain localization and plastic instability.

A modeling methodology based on an internal state variable model has shown to be particularly useful in analyzing irradiation-induced hardening. A differential form of the evolution law is useful in practice, since the irradiation history can be easily integrated to predict the extent of damage – as characterized by $\hat{\sigma}_y$. This presents a compelling motivation to adopt a state-variable modeling approach. In a variety of materials, Farrell and Byun[17,4] plotted the yield stress increase due to irradiation versus exposure in dpa on logarithmic coordinates and concluded the low-dose regime is fundamentally different from the high dose regime. It is interesting that Equation 12 appears to provide a possible fit to the data (e.g., the 9Cr-1MoVNb (HFIR) data plotted in Figure 4 of Reference 17) without need for a transition in mechanism. Kocks[18] and Estrin and Mecking[19] have given physical significance to the Voce law through a competition between dislocation accumulation and annihilation processes. The application of this to the case of irradiation damage should be further pursued.

It is interesting that evolution of the irradiation-induced damage can be described using a Voce law, but more work is required to better define the behavior and to understand differences between data sets. A systematic experimental investigation of the potential effect of exposure temperature on evolution should be undertaken. This could entail, for example, exposure at 50°C, 150°C, and 250°C (if feasible) to levels from 0.5 dpa (or perhaps even lower) to as high as 10 dpa. Measurement and analysis of stress-strain curves where a single parameter is varied (e.g., dpa at a constant exposure temperature) would allow a clearer assessment of the potential effect of exposure temperature. Theoretical modeling might be instructive here as well.

The number of model parameters that comprise the MTS model is often considered to be an issue, and can lead to the conclusion that almost any data set can be fit. Indeed, there are many more model parameters in the equations used to make the model predictions in Figures 3 and 5 than there are data points. However, many of the model parameters are ineffective at yielding discernable differences in the fits. For example, selection of p's, q's, and $\dot{\varepsilon}_o$'s in Equations 3 and 4 have no effect. As has been discussed in previous papers, these parameters allow for a more realistic obstacle profile[20], but their effect is subtle – at best. With huge data sets – such as the data sets in references 6,7, and 8 – a slight improvement in the data fit was obtained with certain values of these coefficients. Therefore, the choice of p's, q's, and $\dot{\varepsilon}_o$'s has been motivated by this original work, and no effort to change these parameters has been made. The important model parameters are those that have been listed in Tables 1, II, IV, and V. Throughout this paper, the practice has been to choose parameters and hold them constant unless a justification exists – e.g., in evolution of $\hat{\sigma}_v$ or $\hat{\sigma}_\varepsilon$.

CONCLUSIONS

Irradiation damage measurements in mild steel and 316 SS were analyzed according to a state variable constitutive model. The defect population is characterized as one that is thermally activated – meaning dislocation movement through the defected structure depends on temperature and strain rate, even though the dependence is weak. Addition of a mechanical threshold stress – $\hat{\sigma}_v$ – to the standard MTS model enabled agreement between the yield stresses and full stress-strain curves measured in irradiated material. Evolution of $\hat{\sigma}_v$ with dose was observed to follow a Voce law similar to the evolution of $\hat{\sigma}_\varepsilon$ with strain in the MTS model. Agreement between measured and predicted stress strain curves required a modification of the latter (Equation 10 versus Equation 7). Even though some of the published measurements involved exposures at varying temperatures, it was not possible to assign temperature dependence to the Voce law for $\hat{\sigma}_v$. Suggestions were made for further measurements and, perhaps, theoretical modeling to enable a more definitive assessment of this. A state-variable model for irradiation damage appears to be feasible and to offer benefits in predictive capability.

[1] S. J. Zinkle and B. Į. Singh, Microstructure of Įeutron-irradiated Iron Before and After Tensile Deformation, TMS Symposium on Microstructural Processes in Irradiated Materials, San Francisco, CA (February, 2005).

[2] L. K. Mansur, A. F. Rowcliffe, R. K. Įanstad, S. J. Zinkle, W. R. Corwin, R. E. Stoller, TITLE, J. Įuclear Mater. **329-333**, 166 (2004)

[3] T. S. Byun and K. Farrell, Irradiation Hardening Behavior of Polycrystalline Metals After Low Temperature Irradiation, J. Įuclear Mater. **326**, 86-96 (2004)

[4] T. S. Byun and K. Farrell, Plastic Instability in Polycrystalline Metals After Low Temperature Irradiation, Acta Mater. **52**, 1597-1608 (2004)

[5] M. Meshii and J. W. Kauffman, Quenching Studies on Mechanical Properties of Pure Gold, Acta Metal. **7**, 180-186 (1959)

[6] P. S. Follansbee and U. F. Kocks, A Constitutive Description of the Deformation of Copper Based on the use of the Mechanical Threshold Stress as an Internal State Variable, Acta Metall. **36**, 81 (1988)

[7] P. S. Follansbee, J.C. Huang, and G. T. Gray III, Low Temperature and High-Strain-Rate Deformation of Įickel and Įickel-Carbon Alloys and Analysis of the Constitutive Behavior According to an Internal State Variable Model, Acta Metall. **38**, 1241 (1990)

[8] P.S. Follansbee and G. T. Gray III, An Analysis of the Low Temperature, Low and High Strain-Rate Deformation of Ti-6Al-4V, Metall. Trans. A, **20A**, 863 (1989)

[9] A. Kelly and R. B. Įicholson, Strengtheni ng Methods in Crystals, Halstead Press, ĮY, 41 (1971)

[10] K. L. Murty, Is Įeutron Radia tion Exposure Always Detrimental to Metals (Steels)?, Letters to Įature **308**, 51-52 (1984)

[11] C. Albertini and M. Montagnani, Dynamic Uniaxial and Biaxial Stress-Strain Relationships for Austenitic Stainless Steels, Įuclear Engnr. and Design **57**, 107-123 (1980)

[12] T. S. Byun, K. Farrel, E. H. Lee, L. K. Mansur, S. A. Maloy, M. R. James, and W. R. Johnson, Temperature Effects on the Mechanical Properties of Candidate SĮ S Target Container Materials after Proton and Įeutron Irradiation, Oa k Ridge Įational Laboratory, SĮS-101060100-TR0003-R00, (2001)

[13] U. Dai, G. W. Egeland, and B. Long Tensile Properties of EC316LĮ Irradiated in SIĮQ to 20 dpa", J. of Įuclear Mater. **377**, 109-114 (2008).

[14] P. S. Follansbee, Analysis of Deformation Kinetics in Seven BCC Pure Metals Using a Two Obstacle Model, Mater. and Metall. Trans. A, in press (2010)

[15] P. S. Follansbee, A Lower Bound Strength Model for AISI 304 SS, Proceedings 2010 MS&T Conference, Houston, TX (2010)

[16] Steichen, J. M. and M. M. Paxton, Interim Report – Effect of Strain Rate on the Mechanical Properties of Austenitic Stainless Steels, Hanford Engineering Development Laboratory, HEDL-TME-71-56, May, 1971.

[17] K. Farrell and T. S. Byun, Tensile Properties of Ferritic/Martensitic Steels irradiated in HFIR, and Comparison with Spallation Irradiation Data, J. Įuclear Matl. **318**, 274-283 (2003)

[18] U. F. Kocks, J. Engng. Mater. Tech. **98**, 76 (1976)

[19] Y. Estrin and H. Mecking, Acta Metall. **32**, 57 (1976)

[20] U. F. Kocks, A. S. Argon, and M. F. Ashby, Thermodynamics and Kinetics of Slip, Progress of Materials Science Vo. 19, Pergamon Press, ĮY, 142 (1975)

RADIATION SHIELDING SIMULATION FOR WOLLASTONITE-BASED CHEMICALLY BONDED PHOSPHATE CERAMICS

J. Pleitt[1], H. A. Colorado[2, 3], C. H. Castano[1*]

[1]Missouri University of Science and Technology, Nuclear Engineering Department.
[2]Materials Science and Engineering, University of California, Los Angeles.
[3]Universidad de Antioquia, Mechanical Engineering Department. Medellin-Colombia.

*Corresponding author: castanoc@mst.edu.

ABSTRACT
   Neutron and gamma attenuation on Wollastonite-based Chemically Bonded Phosphate Ceramics (Wo-CBPCs) is studied with Monte Carlo simulation (MonteCarlo N-Particle, MCNP). Wollastonite-based CBPC is a composite material with several crystalline (Wollastonite and brushite) and amorphous phases (silica and amorphous calcium phosphates). The effect of lead incorporated into the ceramic as a gamma shield has been examined across various gamma energies as well as neutron attenuation of the ceramic. Besides this simulation effort, neutron activation analysis was also performed experimentally at the Missouri University of Science and Technology Reactor (MSTR) to determine the activation by neutrons of the ceramic. In the past, it was found that the attenuation coefficient was substantially improved by the incorporation of Pb in the Wo-CBPC. The improvements at energies above 300 keV ranged from 32 to 193.8% [1]. In the present simulation, we found that the addition of PbO increased the neutron attenuation by 180 to 400% and accurately predicted gamma attenuation values for photon energies above 1000 keV.

INTRODUCTION
   Chemically Bonded Ceramics (CPCs) are inorganic solids generated by chemical reactions at low temperatures. This method avoids high temperature processing which allows the ceramics to be processed inexpensively in high volume production. The ceramics have also been shown to have high thermal shock resistance and have been used in firewall applications[2], nuclear waste and encapsulation[3], electrical materials[4], as well as composite materials[5]. This research focuses on the simulation of previous formulations with MCNP simulations to evaluate the effectiveness of the experiment in determining linear attenuation coefficients and to determine how to optimize the neutron attenuation values. Various models have already been used to determine shielding properties of concrete with additives[6,7], as well as designing optimal shielding properties of concrete ratios[8]. Also there has been studies examining both gamma ray and neutron attenuation of composite materials in MCNP simulations as well as optimization studies on ceramic samples to improve their shielding properties[9,10]. The linear attenuation value is an important parameter to calculate the maximum amount of radiation released from a shielded source. In order to make sure that the public remains below their dose limit an analysis of linear attenuation coefficients for gamma with both experimental data and MCNP simulations was performed as well as MCNP simulations for neutron attenuation.
   In a previous paper it was found that incorporating PbO into CBPC improved the linear attenuation coefficient for gammas from 32 to 193.8%[1] at different energies. The current simulations will help us improve the formulation for both gamma and neutrons should other additives be incorporated in the structure.

EXPERIMENTAL DETAILS

In a previous study an experiment was performed using a calibrated Europium Source that contained Europium-152, Europium 154, and Europium-155 which provided a wide range of energies to test for Linear Attenuation coefficients. Due to decay of the Europium source only 7 energy values were evaluated due to the clarity of their peaks. A high purity Germanium Detector was used to determine the energy values of the peaks. The lead collimators were placed to generate a gamma beam and can be seen in Figure 1 for both the theoretical set-up and the actual set-up.

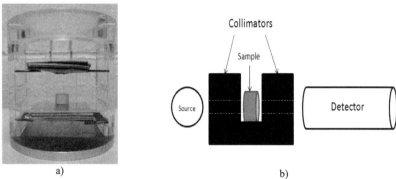

a)   b)

Figure 1. a) experimental set-up for Gamma Rays, b) representation

The linear attenuation coefficient is determined from equation 1 were I_o is initial flux, I is flux with sample included, x is the thickness of the sample in cm, and μ is the attenuation coefficient in cm⁻¹.

$$I = I_o e^{-\mu x} \tag{1}$$

Using this equation one can determine the linear attenuation coefficient by solving for μ in cm⁻¹. One thing to note about this is that this equation is energy dependent, different energies result in different linear attenuation values and therefore a graph of μ versus energy or tabled values are regularly used to present the information. Using these values one can determine the necessary thickness of shielding for a variety of radiation energy sources.

The energy spectrum for the calibrated Europium source is shown in Figure 1c.

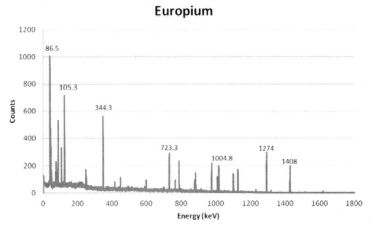

Figure 1. c) Europium Energy Spectrum

Using the Europium spectrum one can obtain many different energy peaks that can be used to determine a linear attenuation graph for various materials. Once this graph is constructed one can use equation 1 to determine the minimum thickness of a radiation shield to achieve the necessary requirements for dose limits. In order to determine the peaks the results from the experiment integrated from the peak height across 10 energy channels with an average width of 0.1 keV.

SIMULATION DETAILS

MCNPX was used for simulating gamma attenuation by the CBPCs. MCNPX is a general purpose Monte Carlo transport code that can track thousands of particles over large energy ranges(0 to 150MeV)[11] and has been benchmarked for various uses[12]. The simulation used the same basic set-up as the experiments, however the source was approximated as a point source due to the difficulty in determining the exact location of the various europium particles in the source. The samples sed were 2 cm thick samples of the same cylindrical shape of the manufactured samples. The samples were estimated to be homogeneously mixed. In order to add porosity of the substance into the simulation, the calculated porosity of the sample was used to add in an air pockets to the sample, however for calculating the linear attenuation the initial thickness of the sample was used. The model for CBPC with and without PbO added is summarized in Table I.

Table I. MCNPX inputs

Height (cm)	Porosity (%)	Modeled Height (cm)	Density (g/cm^3)	Diameter (cm)	Atomic Composition (%)
2	21.76	1.56484	2.112	1.1	47.4- O 36.8- H 5.2- Si 5.2- Ca 5.2 -P
2	21.76	1.56484 cm	2.667	1.1	41.2- O 32.6- H 4.6- Si 4.6- Ca 4.6 -P 11- Pb

For the neutron simulations a similar setup as the measurements for the gamma ray detection was used. A neutron beam of varying energies from thermal to fast (1/40eV to 2.5 MeV) was used to determine the attenuation coefficient for the ceramic. Since reactors are usually used to determine neutron shielding details[13] and we plan to measure the neutron attenuation using a Pu-Be neutron source we studied the shielding capabilities at energies up to 2.5 MeV. The same compositions as well as the air pocket addition for porosity were included in the simulations.

The model was examined for gamma attenuation energies for a Europium source containing 7 energy peaks at 86.5, 105.3, 344.3, 723.3, 1004.8, 1274, and 1408 keV. These give a wide spectrum of energies useful for obtaining a linear attenuation graph. For the 7 neutron energies chosen ranging from thermal energies to 2.5MeV, 500 million particles were run in order to reduce the error of the MCNP simulation for each run.

ANALYSIS

The experiment used a Europium Source to determine the linear attenuation coefficient for Chemically Bonded Phosphate Ceramics (CBPCs). The original experiment was also examined in MCNPX to verify the results of the measurement[1]. The measured porosity of the sample was included in the calculation as additional uniformly distributed air pockets. The results measured the flux over 10 energy channels averaging 0.1 keV for each of the initial energies from Europium to obtain the uncollided flux of the particles. The results from both MCNP and the actual measured values are plotted in Figure 2 for CBPC without lead added.

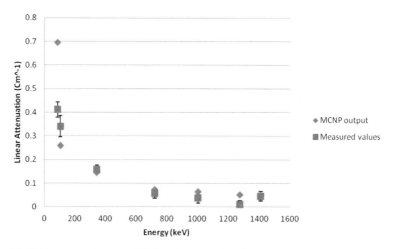

Figure 2. MCNP output and measured Values for a Europium Source and CBPCs with no PbO content

The simulations were performed with MCNPX evaluating the energy peaks within a 10kev range of the initial energy. Porosity of the substances was included to obtain results that were similar to measured results for the Europium gamma source used. For most values the MCNPX output was slightly higher than the measured value which is likely due to the porosity of the CBPCssince a change of just 1% porosity causes a 5% change in attenuation values. For the lower energy values (E<XXX keV) values are close to the Compton edge which could cause differ notably from the calculated values.

The error for the simulations was around 0.025% for all of the simulated values. For 1274.5 keV there is a large difference between the simulated and measured values. This had been noted in the previous study that the data point obtained appeared to be an outlier[1] causing the discrepancy between simulated and actual values. The simulations were also performed for CBPCs with PbO added and are shown in Figure 3.

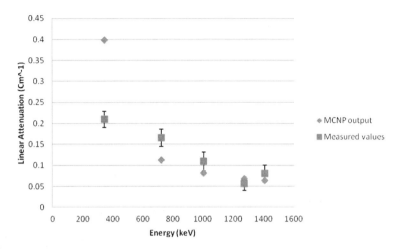

Figure 3. MCNP output and Experimental measured values with 50 wt% PbO added in

One thing to note is that there is only five gamma peaks in this sample set as compared to the content the CBPC without PbO added in. The lower energy peaks (XX keV and XX keV) are missing since they had been reduced to background and a gamma peak could not be determined in the low energy ranges thereby removing the data points. The MCNP outputs for 50 wt% PbO mixture added into the ceramic tended to be lower than the actual measured values of the ceramic excepting the 1274.5keV and 344.3keV energy peaks. For the 344.3keV the likely cause is due to the scattering of higher energy peaks since the initial calculation required over 20 energy channels to be integrated together to obtain the peaks as well as baseline subtraction to eliminate background noise, this would result in a measured value that is lower than its actual for lower energy gammas. At 1274.5keV the energy values are within one standard deviation of each other which suggest that the simulation is accurately predicting the measured results. For the most part as the energy increased the differences between the simulations predictions and the measured results became smaller which suggests that there were errors in the simulation of the linear attenuation coefficients likely due to differences in the porosity of the simulated CBPCs and the experimental. These errors are likely due to accidental beam adjustment as well as scattering from higher energy gamma rays. The scattering error cannot be adjusted for, however making sure that the gamma source remains stationary in between sample loadings would improve the accuracy of these measurements.

For the neutron attenuation test with MCNP varying energies ranging from thermal to fast were used to determine the attenuation coefficients for neutrons. The energy spectrum for CBPC without any PbO and as well as the spectrum with PbO included is shown in Figure 4.

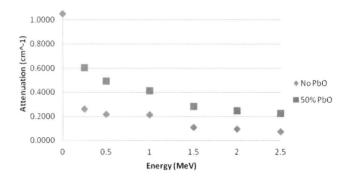

Figure 4. Neutron Linear Attenuation for CBPC

As can be seen CPBC attenuates neutrons very well with a value above 1 for thermal energy neutrons. When PbO was added to the system, the thermal energy attenuation increased to 4.8 cm^{-1} increasing by about 400%. At the higher energy values for fast neutrons CPBC performs decently with most values between 0.2 and 0.3 cm^{-1}. The addition of PbO caused an increase ranging from 180 to 300% for the attenuation values.

For The 50 wt% PbO there was a 458% increase in the attenuation of thermal neutrons and around a 200% increase in the attenuation of fast neutrons. This increase corresponds similarly with the increase of the gamma linear attenuation values. PbO values from 0.5 to 2.5 MeV all exist in the resonance region for Pb, it is likely that this region is causing issues with the output (sudden changes in attenuation values). For the other major elements (Si, Ca, O, and P) they are likewise in the resonance region however due to the lower cross-sections for these elements there is no effect observed. The likely cause for the increase in attenuation is due to the increased density of hydrogen which has the highest cross section for the elements in this region which resulted in improved neutron attenuation.

CONCLUSIONS

From the calculations and validation experiment performed it is clear that MCNP can be used to predict linear attenuation coefficients for ceramics so long as the porosity and density are well known and taken into account correctly. It is believed that the variations in the values of the coefficients is caused by errors in both the experiment and the porosity added into the simulation. With just 1% increase in the porosity the attenuation decreased by 5%. Other than low energy outliersoutliers due to inherent uncertainty in the experimental measurements MCNP can arrive at close (20% or lower) to the experimental values for linear attenuation coefficients.

For the gamma attenuation simulations, MCNP arrived within 0.02 cm^{-1} to the measured values for the higher energy gammas and was off anywhere from 0.1 cm^{-1} to 0.3 cm^{-1}. in predicting low energy values. This is likely due to issues that were created by the experimental procedure and apparently not due to issues with the simulation. When PbO was added there were some issues in predicting the values obtained from the experiment for energies below 1000 keV. This is also likely due to inherent uncertainty in the measurements.

The increase in the neutron attenuation for CBPC is likely due to the increase in the density and total composition of hydrogen in the simulations. Since Hydrogen has a very large cross section compared to the other elements in the CBPC the overall increase in the total attenuation is due to the increase in the overall density of hydrogen in the sample. The addition of PbO to CBPC showed a substantial increase in both measurements of the attenuation coefficient for both neutrons and gamma shielding purposes.

The results from the research show promising results for the improvement of shields made of CBPCs for both gamma and neutron attenuation with the addition of additives. By adding other additives and observing the changes in shielding properties it is hoped that the shielding capabilities of this ceramic can be optimized even further.

ACKNOWLEDGMENTS

The authors desire to express their gratitude to Colciencias from Colombia for support Henry A. Colorado. Also, to the staff of the Missouri S&T nuclear reactor Bill Bonzer, Craig Reisner, and Thorne Kontos for their help setting up the shielding experiments

REFERENCES

1. H. A. Colorado, C. Hiel and H. T. Hahn, J. M. Yang, J. Pleitt, C. Castano. Wollastonite-based Chemical Bonded Phosphate Ceramic with lead oxide contents under gamma irradiation. Journal of Nuclear Materials, 2011, DOI: 10.1016/j.jnucmat.2011.08.043.

2. H. A. Colorado, C. Hiel and H. T. Hahn. Chemically Bonded Phosphate Ceramics composites reinforced with graphite nanoplatelets. Composites Part A. Composites: Part A 42 (2011) 376–384.

3. D. Singh, S. Y. Jeong, K. Dwyer and T. Abesadze. Ceramicrete: a novel ceramic packaging system for spent-fuel transport and storage. Argonne National Laboratory.Proceedings of Waste Management 2K Conference, Tucson, AZ, 2000.

4. J. F. Young and S. Dimitry. Electrical properties of chemical bonded ceramic insulators. J. Am. Ceram. Soc., 73, 9, 2775-78 (1990).

5. T. L. Laufenberg, M. Aro, A. Wagh, J. E. Winandy, P. Donahue, S. Weitner and J. Aue. Phosphate-bonded ceramic-wood composites. Ninth International Conference on Inorganic bonded composite materials (2004).

6. E. Yilmaz, H. Baltas, E. Kiris, I. Usabas, U. Cevik, A.M. El-Khayatt. Gamma ray and neutron shielding properties of some concrete materials. Annals of Nuclear Energy (2011) 2204-2212

7. A.M. El-Khayatt, Radiation shielding of concretes containing different lime silica ratios. Annals of Nuclear Energy, 2010.

8. Elbio Calzada, Florian Grunauer, Burkhard Schillinger, Harald Turck, Reusable shielding material for neutron- and gamma- radiation. Nuclear Instruments and Methods in Physics Research A, 2011.

9. Majid Jalali, Ali Mohammadi. Gamma ray attenuation coefficient measurement for neutron-absorbant materials. Radiation Physics and Chemistry 77 (2008).

10. Huasi Hu, Qunshu Wang, Juan Qin, Yuelei Wu, tiankui Zhang Zhonseng Xie, Xinbiao Jiang, Guoguang Zhang, Hu Xu, Xiangyang Zheng, Jing Zhang, Wenhao Liu, Zhenghong Li, Boping Zhang, Linbo Li, Zhaohui Song, Xiaoping Ouyang, Jun Zhu, Yaolin Zhao, Xiaoqin Mi, Zhengping Dong, Cheng Li, Zhenyu Jiang, and Yuanpin Zhan. Study on Composite Material for Shielding Mixed Neutron and γ-rays. IEEE Transactions on Nuclear Science, Vol 55 (2008).

11. Denise B. Pelowitz, MCNPX Users Manual ver 2.6.0.Los Alamos National Laboratory, (April 30, 2008)

12. Daniel J. Whalen, David A. Cardon, Jennifer L. Uhle, John S. Hendricks. MCNP: Neutron Benchmark Problems. Los Alamos National Laboratory, 1991.

13. A.S. Makarious, I. I. Bashter, A. El-Asyed Abdo, and W.A. Kansouh. Measurement of Fast Neutrons and Secondary Gamma Rays in Graphite, Annals of Nuclear Energy Vol 23, No. 7 (1996).

EMPIRICAL MODEL FOR FORMULATION OF CRYSTAL-TOLERANT HLW GLASSES

J. Matyáš[1], A. Huckleberry[1], C.A. Rodriguez[1], J.D. Vienna[1], A.A. Kruger[2]
1. Pacific Northwest National Laboratory
   Richland, WA, USA
2. Office of River Protection
   Richland, WA, USA

ABSTRACT
   Historically, high-level waste (HLW) glasses have been formulated with a low liquidus temperature (T_L), or temperature at which the equilibrium fraction of spinel crystals in the melt is below 1 vol % ($T_{0.01}$), nominally below 1050°C. These constraints cannot prevent the accumulation of large spinel crystals in considerably cooler regions (~ 850°C) of the glass discharge riser during melter idling and significantly limit the waste loading, which is reflected in a high volume of waste glass, and would result in high capital, production, and disposal costs. A developed empirical model predicts crystal accumulation in the riser of the melter as a function of concentration of spinel-forming components in glass, and thereby provides guidance in formulating crystal-tolerant glasses that would allow high waste loadings by keeping the spinel crystals small and therefore suspended in the glass.

INTRODUCTION
   The high-level radioactive waste (HLW) from the Hanford and Savannah River Sites is being vitrified in stable borosilicate glass for long-term storage and disposal. This process is time consuming and expensive because it is highly dependent on loading of HLW in glass and on the rate of HLW glass production. The current HLW melters are projected to operate in an inefficient manner as they are subjected to artificial constraints that limit waste loading to far below its intrinsic level.[1] These constraints, such as liquidus temperature (T_L) of glass or the temperature at which the equilibrium fraction of spinel crystals in the melt is below 1 vol % ($T_{0.01}$), nominally below 1050°C, were imposed to prevent clogging of the melter with spinel crystals that can accumulate at the bottom and in the glass discharge riser based upon operational experience with static melters (i.e., non-bubbled).[2]
   To protect the melter from detrimental accumulation of spinel crystals, attention has been focused on studying the settling of spinel crystals in molten glasses[3-6] as well as transparent liquids[7,8]. Lamont and Hrma[4] observed the parabolic shape of the settling front indicating that the settling crystals generated a convective cell within the melt. Klouzek et al.[5] determined that the measured settling distances between the glass level and the uppermost crystals in the centerline of the crucible were less than 10% smaller than the distances calculated with the modified Stoke's law. Matyas et al.[3] determined the accumulation rate of crystals as a function of spinel forming components and noble metals, and revealed a beneficial effect of suppressing the crystal size and accumulation rate through additions of Fe and noble metals. Matlack et al. reported that the high-crystal content glasses of up to 4.2 vol% at 950°C have been successfully discharged from the DuraMelter® DM-100 after about 8 days of melter idling at 950 °C.[6]
   The goal of this work was to develop an empirical linear model of spinel settling that can predict crystal accumulation in the riser as a function of glass composition and therefore provide the guidance to formulate crystal-tolerant glasses for higher waste loading. By keeping the spinel

crystals small and therefore limiting spinel deposition in the melter, these glasses will allow high waste loading without decreasing melter lifetime.

EXPERIMENTAL
Glass Matrix Design and Fabrication

Glass matrix of twelve compositions was developed by changing concentrations of Cr_2O_3, NiO, Fe_2O_3, ZnO, MnO, Al_2O_3, and noble metals (Rh_2O_3 and RuO_2) one or two components-at-a-time from the baseline glass composition (BL) while proportionally decreasing the concentration of all other components. The concentration of these components was varied to encompass their variation in Hanford HLW, see Table I. Table II shows the composition of designed glasses, including the baseline glass.

Table I. Concentration Variation of Noble Metals, Cr, Ni, Fe, Zn, Mn, and Al in Hanford HLW in Mass Fraction of Oxides.

Component	Minimum	Maximum
Rh_2O_3	1.1E-08	0.0004
RuO_2	2.8E-06	0.0024
Cr_2O_3	0.0027	0.0584
NiO	0.0012	0.0351
Fe_2O_3	0.0140	0.5244
ZnO	0.0005	0.0181
MnO	0.0018	0.0668
Al_2O_3	0.0704	0.7350

Glass batches were prepared from AZ-101 simulant[9] and additives (H_3BO_3, Li_2CO_3, Na_2CO_3, and SiO_2). Extra Cr, Ni, Fe, Zn, Mn, and Al were added as Cr_2O_3, NiO, Fe_2O_3, ZnO, MnO, Al_2O_3, and Rh_2O_3. Ruthenium was added in the form of ruthenium nitrosyl nitrate solution drop by drop to 100 g of SiO_2 that was dispersed on a Petri dish. The SiO_2 cake was dried in oven at 105°C for 1 hour, quenched, and hand-mixed in the plastic bag with the rest of the glass batch. Then, the glass batch was milled in an agate mill for 5 min to ensure homogeneity.

Glasses were produced in Pt-10%Rh crucibles following a two-step melting process: 1) melting of homogenized glass batches and 2) melting of produced glasses after quenching and grinding. The melting temperature for Ni1.5/Al12 and Fe20/Ni1.5 glasses was 1250°C and 1300°C, respectively. The other glasses were melted at 1200°C.

Table II. Composition of Designed Glasses in Mass Fraction of Oxides and Halogens.

Component	BL	Cr0.6	Cr1.2	Ni1.07	Ni1.5	Ni1.5/nm[a]	Fe20	Fe20/Ni1.5	Mn1	Mn2.5	Zn0.6	Ni1.5/Al12
Al_2O_3	0.0821	0.0817	0.0813	0.0817	0.0814	0.0814	0.0768	0.0760	0.0816	0.0803	0.0816	**0.1200**
B_2O_3	0.0799	0.0796	0.0791	0.0796	0.0792	0.0792	0.0748	0.0739	0.0794	0.0782	0.0794	0.0758
BaO	0.0009	0.0009	0.0009	0.0009	0.0009	0.0009	0.0008	0.0008	0.0009	0.0009	0.0009	0.0009
CaO	0.0057	0.0057	0.0056	0.0057	0.0057	0.0056	0.0053	0.0053	0.0057	0.0056	0.0057	0.0054
CdO	0.0065	0.0065	0.0064	0.0065	0.0064	0.0064	0.0061	0.0060	0.0065	0.0064	0.0065	0.0062
Cr_2O_3	0.0017	**0.0060**	**0.0120**	0.0017	0.0017	0.0017	0.0016	0.0016	0.0017	0.0017	0.0017	0.0016
F	0.0001	0.0001	0.0001	0.0001	0.0001	0.0001	0.0001	0.0001	0.0001	0.0001	0.0001	0.0001
Fe_2O_3	0.1451	0.1445	0.1436	0.1445	0.1438	0.1438	**0.2000**	**0.2000**	0.1442	0.1420	0.1443	0.1377
K_2O	0.0034	0.0034	0.0034	0.0034	0.0034	0.0034	0.0032	0.0031	0.0034	0.0033	0.0034	0.0032
Li_2O	0.0199	0.0198	0.0197	0.0198	0.0197	0.0197	0.0186	0.0184	0.0198	0.0195	0.0198	0.0189
MgO	0.0013	0.0013	0.0013	0.0013	0.0013	0.0013	0.0012	0.0012	0.0013	0.0013	0.0013	0.0012
MnO	0.0035	0.0035	0.0035	0.0035	0.0035	0.0035	0.0033	0.0032	**0.0100**	**0.0250**	0.0035	0.0033
Na_2O	0.1866	0.1858	0.1847	0.1858	0.1850	0.1849	0.1746	0.1726	0.1854	0.1826	0.1855	0.1771
NiO	0.0064	0.0064	0.0063	**0.0107**	**0.0150**	**0.0150**	0.0060	**0.0150**	0.0064	0.0063	0.0064	**0.0150**
P_2O_5	0.0032	0.0032	0.0032	0.0032	0.0032	0.0032	0.0030	0.0030	0.0032	0.0031	0.0032	0.0030
SiO_2	0.4031	0.4014	0.3989	0.4014	0.3996	0.3995	0.3772	0.3729	0.4005	0.3944	0.4008	0.3825
SO_3	0.0008	0.0008	0.0008	0.0008	0.0008	0.0008	0.0007	0.0007	0.0008	0.0008	0.0008	0.0008
TiO_2	0.0003	0.0003	0.0003	0.0003	0.0003	0.0003	0.0003	0.0003	0.0003	0.0003	0.0003	0.0003
ZnO	0.0002	0.0002	0.0002	0.0002	0.0002	0.0002	0.0002	0.0002	0.0002	0.0002	**0.0060**	0.0002
ZrO_2	0.0416	0.0414	0.0412	0.0414	0.0412	0.0412	0.0389	0.0385	0.0413	0.0407	0.0414	0.0395
Cl	0.0002	0.0002	0.0002	0.0002	0.0002	0.0002	0.0002	0.0002	0.0002	0.0002	0.0002	0.0002
Ce_2O_3	0.0020	0.0020	0.0020	0.0020	0.0020	0.0020	0.0019	0.0019	0.0020	0.0020	0.0020	0.0019
CoO	0.0001	0.0001	0.0001	0.0001	0.0001	0.0001	0.0001	0.0001	0.0001	0.0001	0.0001	0.0001
CuO	0.0004	0.0004	0.0004	0.0004	0.0004	0.0004	0.0004	0.0004	0.0004	0.0004	0.0004	0.0004
La_2O_3	0.0022	0.0022	0.0022	0.0022	0.0022	0.0022	0.0021	0.0020	0.0022	0.0022	0.0022	0.0021
Nd_2O_3	0.0018	0.0018	0.0018	0.0018	0.0018	0.0018	0.0017	0.0017	0.0018	0.0018	0.0018	0.0017
SnO_2	0.0010	0.0010	0.0010	0.0010	0.0010	0.0010	0.0009	0.0009	0.0010	0.0010	0.0010	0.0009
Total	1.0000	1.0000	1.0000	1.0000	1.0000	0.9997	1.0000	1.0000	1.0000	1.0000	1.0000	1.0000

[a] Added 0.0003 Rh_2O_3 and 2.9E-5 RuO_2

Settling experiments

The double crucible test was used to study the accumulation of spinel crystals.[3,8] The alumina crucible was nested in the big silica crucible, held in place with the core-drilled silica crucible, and covered with molten glass to eliminate the Marangoni convection in the meniscus and bubble generation at the bottom of silica crucibles. First, glass powders were melted in Pt-10%Rh crucible at 1200°C for 1 h to dissolve spinel crystals that might formed during the quenching of the glass. Then, the crucible was removed from the melting furnace and molten glass was poured into three double crucibles that were rested inside the furnace at 850°C, mimicking the temperature in the glass discharge riser. The crucibles were removed at various times and cross-sectioned. The rectangular pieces 3 cm wide and 5 cm long were cut out from the bottom of the crucibles, thin-sectioned, and analyzed with scanning electron microscopy-energy dispersive spectroscopy (SEM-EDS) and Clemex image analysis to determine the thickness of the spinel sludge layer.

Empirical Model of Spinel Crystal Settling
    Three stages were identified during the settling experiments in the double crucibles: 1) latency period with no settling, 2) settling period with constant settling rate of spinel, and 3) end of settling period with a low and gradually decreasing settling rate of spinel due to a smaller and smaller number of settling crystals. Only the sludge layer thickness data for glasses that were collected during the constant settling rate period were used to build an experimental model predicting crystal accumulation in the glass discharge riser as a function of seven major components (Al_2O_3, Cr_2O_3, Fe_2O_3, ZnO, MnO, NiO, and Others). The constant settling rate allowed us to use a general linear model in the form:

$$h = \sum_{i=1}^{7} h_i x_i + t \sum_{i=1}^{7} s_i x_i \qquad (1)$$

where h is the layer thickness (μm), h_i is the compositional dependent intercept coefficient (μm), x_i is the i-th component mass fraction, t is the settling time (h), and s_i is the compositional dependent velocity coefficient (μm/h).

RESULTS AND DISCUSSION
    Table III shows the calculated coefficients h_i and s_i, R^2 (expresses the fraction of the variability accounted for by the model), and R^2_{adj} (adjust R^2 for the number of parameters used in fitting the model). Negative coefficients s_i for Al_2O_3 and Fe_2O_3 suggest that these components decrease the settling rate of crystals. In contrast, additions of MnO, ZnO, Cr_2O_3, and NiO to the baseline glass increase the settling rate. Nickel oxide stands out as the most troublesome component because of the formation of large spinel crystals with a more than six times faster accumulation rate than the crystals, e.g., in Cr-rich glass. The detrimental effect of this component on the settling rate can be significantly suppressed by introducing the noble metals or Fe_2O_3 to the glass. The negative coefficients h_i for MnO, ZnO, Cr_2O_3, and NiO only indicate, but do not predict, the length of the latency period. This period is dependent on the initial growth rate of crystals to the size at which crystals start to settle.

Table III. Component Coefficients Calculated with PNNL Model

Components	h_i (μm)	s_i (μm/h)
Al_2O_3	8816.97	-350.41
Fe_2O_3	4304.182	-49.9117
MnO	-7498.52	259.3812
ZnO	-12257.6	313.0436
Cr_2O_3	-40257.3	443.5807
NiO	-197477	2672.734
Others	-366.91	27.00287
R^2	0.985	
R^2 adj	0.975	

Figure 1 shows the predictive versus measured thicknesses of a spinel sludge layer for tested glasses. The linear empirical model with coefficients h_i and s_i expressed as a linear function of mass fractions of seven major components fits the 35 data points reasonably well, $R^2=0.985$, and can become an efficient tool to formulate the crystal-tolerant glasses that would ultimately allow a substantial increase in the waste loading.

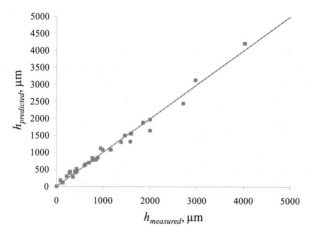

Figure 1. Predicted vs. Measured Spinel Layer Thickness.

CONCLUSION

The developed 7-component model can predict reasonably well the crystal accumulation in the riser as a function of glass composition and therefore allows higher waste loadings, and at the same time protects the HLW glass melter from detrimental accumulation of spinel in the glass discharge riser during melter idling. In the future work, we plan to expand the compositional region covered by our model, and thereby improve its predictive performance. We will also elucidate the accumulation rate of spinel crystals at temperatures above 850°C because the temperature in the glass discharge riser varies and can exceed 950°C during melter idling. Additionally, we will investigate the impact of different components on agglomeration of particles and on the shape, size, and concentration of crystals.

ACKNOWLEDGEMENT

This work was funded by the U.S. Department of Energy's Environmental Management Program EM20. Pacific Northwest National Laboratory is operated by Battelle for the U.S. Department of Energy under Contract DE-AC05-76RL01830.

REFERENCES
[1]D.S. Kim and J.D. Vienna, Influence of Glass Property Restrictions on Hanford Glass Volume, *Ceramic Transactions* 132, 105-115 (2002).

[2]P. Hrma, J. Matyáš, and D.S. Kim, Evaluation of Crystallinity Constraint for HLW Glass Processing, *Ceramic Transactions* 143, 133-140 (2003).

[3]J. Matyáš, J.D. Vienna, A. Kimura, M. Schaible, and R.M.Tate, Development of Crystal-Tolerant Waste Glasses, *Ceramic Transactions* 222, 41-51 (2010).

[4]J. LaMont and P. Hrma, A Crucible Study of Spinel Settling in a High-Level Waste Glass, *Ceramic Transactions* 87, 343-348 (1998).

[5]J. Kloužek, J. Alton, T.J. Plaisted, and P. Hrma, Crucible Study of Spinel Settling in High-Level Waste Glass, *Ceramic Transactions* 119, 301-308 (2001).

[6]K.S. Matlack, W.K. Kot, W. Gong, W. Lutze, I.L. Pegg, and I. Joseph, Effects of High Spinel and Chromium Oxide Crystal Contents on Simulated HLW Vitrification in DM100 Melter Tests, Department of Energy Office of River Protection, VSL-09R1520-1 (2009).

[7]J. Matyáš, J.D. Vienna, and M.J. Schaible, Determination of Stokes Shape Factor for Single Particles and Agglomerates, *Ceramic Transactions* 227, 195-203 (2011).

[8]J. Matyáš, J.D. Vienna, M.J. Schaible, C.P. Rodriquez, and J.V. Crum, Development of Crystal-Tolerant High-Level Waste Glasses, Pacific Northwest National Laboratory, Richland, Washington, PNNL-20072 (2010).

[9]R.E. Eibling, R.F. Schumacher, and E.K. Hansen, Development of Simulants to Support Mixing Tests for High Level Waste and Low Activity Waste, Westinghouse Savannah River Company, Savannah River Site, Aiken, SC, SRT-RPP-2003-00098, REV. 0. (2003).

Energy
Conversion/Fuel Cells

NOVEL SOFC PROCESSING TECHNIQUES EMPLOYING PRINTED MATERIALS

P. Khatri-Chhetri, A. Datar, and D. Cormier
Rochester Institute of Technology
Rochester, NY USA

ABSTRACT

Digital printing technologies including inkjet and electrophotography have long been used to produce paper documents. More recently, they have been adapted for use as manufacturing processes to print functional materials and devices. Digital printing processes are attracting interest in great part because they are able to selectively deposit multiple materials in any desired ratio at any location on a substrate. This opens up enormous possibilities for novel material systems and improved functional performance. An added benefit is that printing technologies are inherently scalable for high volume production. This paper describes research in which several different printing technologies are being used to fabricate multi-functional heterogeneous solid oxide fuel cell layers in which the composition and structure are locally controlled.

INTRODUCTION

Solid oxide fuel cells (SOFC's) are traditionally synthesized using one or more thick or thin film processing techniques. Common thick film approaches include tape casting[1], tape calendaring, screen printing[2] and wet spraying[3] where fabrication is usually carried out in layers. Deposition is then followed by firing at high temperatures (1300°C-1700°C). Thin film deposition approaches that have been used to form cell components include chemical and physical process such as chemical vapor deposition (CVD), plasma spraying (PS)[4], spray pyrolysis and spin coating[5].

Reifsnider et al.[6] describe heterogeneous materials as "material systems that consist of multiple materials combined at multiple scales (from nano-to macro-) that actively interact during their functional history in a manner that controls their collective performance as a system at the global level". Conventional synthesis techniques are not particularly well suited for precise local control of chemical composition and/or structure required of heterogeneous materials being developed in conjunction with this effort.

The research described here employs direct write printing techniques to allow chemical composition and structure of an SOFC to be precisely engineered. This allows one to locally control the size and shape of pores in the anode and cathode layers and hence the resulting porosity. The methods reported here are forms of additive manufacturing that utilize micro-extrusion technologies. These methods allow one to control the material deposition rate and path simultaneously. This is a major advantage of this process over various more traditional fabrication processes. The direct-write technique is also capable of grading porosity/composition within a layer in addition to grading porosity in successive layers.

DIRECT-WRITE MATERIAL PRINTING

For this research, an nScrypt Tabletop series microdispensing system with a SmartPump has been utilized for fabrication (Figure 1Figure). The fuel cell material is introduced to the SmartPump in slurry form (referred to as paste or ink), and the tool is capable of printing highly viscous materials up to 1 Million cP. For fabrication, component geometry can be imported from Computer Aided Design (CAD) software, such as ProEngineer or Solidworks. SOFC's can easily be built layer-by-layer from the bottom up. Any pattern can be printed in the XY plane or raised into the Z plane for a conformal surface printing[7].

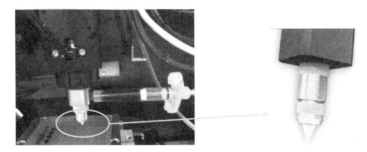

Figure 1. a) nScrypt SmartPump microdispensing tool

The SmartPump is a precise micro-dispensing tool that consists of a positive pressure pump with a computer-controlled needle valve. Air pressure applied to a 3cc syringe barrel containing the SOFC paste can be computer controlled within a range of 0 – 100 psi. The pressure pushes the ink into the main valve body. A servo-controlled needle valve is opened or closed as desired to regulate the flow of ink being pushed into the ceramic nozzle by the air pressure. Ceramic nozzles are available with diameters ranging from 12.5 – 150 μm. The SmartPump can control starts and stops as well as material flow rates for a large range of ink viscosities.

The controllable tool process parameters of an nScrypt Smartpump are:

- Pressure (PSI) – The amount of air pressure applied to syringe containing the ink or paste being dispensed.
- Translation Speed (mm/sec) – The printing speed in x-direction and y-direction.
- Valve Open Position (mm) – The distance the valve opens to control the flow of ink
- Dispensing height (mm) – The distance between the tip of the nozzle and the substrate

Additional factors the user has control over include:

- Nozzle size – The diameter of the orifice in the dispensing tip
- Ink formulation - Particle size, solid loading fraction, viscosity, etc.
- Post deposition sintering schedule – Ramp up rate, hold temperature, hold duration

Interactions between process parameters such as air pressure, translation speed, and dispensing height are highly complex. Likewise, ink rheology varies greatly from one material system to another based on the solvent system, particle size distribution, particle shape, etc. As a general rule, process parameters must be carefully optimized for each new ink formulation developed. Figure 2 illustrates variations in line characteristics for a YSZ ink when a single process parameter (translation speed) is modified.

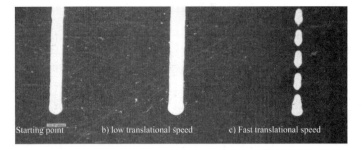

Figure 2. Influence of translation speed on line properties

GRADED POROSITY VIA DIRECT-WRITE MICROEXTRUSION

It has been shown that increasing levels of graded porosity can significantly influence cell performance[8,9,10,11]. The grading of electrodes reduces the mass transfer losses and increases the volumetric reactive surface area close to the electrode-electrolyte interface. However, the highly graded structures produced are complex, and hence are more costly to fabricate. Also, the sintering temperature and fabrication techniques have effects on the microstructures of the graded electrodes and therefore on their electrochemical performance.

In order to demonstrate the suitability of direct-write microextrusion for producing SOFC electrodes having locally graded porosity, NiO-YSZ inks were prepared by mixing nine batches of inks in the proportions shown in Table I. The NiO-YSZ powders (Fuel Cell Materials) were mixed with terpineol-based ink vehicle (Fuel Cell Materials) in a Thinky ARM 310 mixer at 2000 rev/min for 10 minutes. Each of the nine ink formulation shown in Table I were printed on a sintered YSZ substrate to an approximate film thickness of 40 μm using the nScrypt SmartPump micro-dispensing tool. Samples were sintered at 1400°C for 2 hours using a ramp rate of 5°C/minute. The samples were then examined under a Hirox optical microscope in order to qualitatively assess porosity. Micrographs for each experimental condition are shown in Figure 3.

Table I. Summary of various anode compositions to create structures of different porosity

Sample	Total powder weight(g)	%pore former (wt.)	Terpineol (ml)	NiO-YSZ (g)	NiO-YSZ Density (g/cc)	NiO-YSZ (cc)	GRAPHITE Graphite (g)	GRAPHITE Density (g/cc)	GRAPHITE Graphite (cc)	%Pore former (vol.)
1	5.00	0	5.00	5.00	1.55	3.22	0.00	0.13	0.00	0.000
2	5.00	1	5.00	4.95	1.55	3.19	0.05	0.13	0.37	10.50
3	5.00	2	5.00	4.90	1.55	3.15	0.10	0.13	0.75	19.16
4	5.00	3	5.00	4.85	1.55	3.12	0.15	0.13	1.12	26.43
5	5.00	4	5.00	4.80	1.55	3.09	0.20	0.13	1.50	32.61
6	5.00	5	5.00	4.75	1.55	3.06	0.25	0.13	1.87	37.94
7	5.00	10	5.00	4.50	1.55	2.90	0.50	0.13	3.74	56.34
8	5.00	15	5.00	4.25	1.55	2.74	0.75	0.13	5.61	67.21
9	5.00	20	6.00	4.00	1.55	2.57	1.00	0.13	7.48	74.38

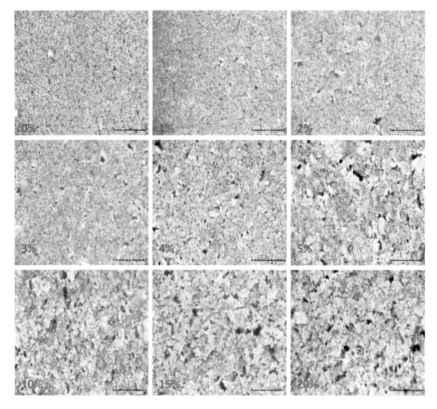

Figure 3. Images of various anodes with different pore former fractions. The number labeled is the %weight of pore-former.

From Figure 3, it can be seen that there is an obvious gradual increase in porosity from 0%wt. to 5%wt. (0% vol. to 33% vol.) graphite. The 5%, 10%, 15% and 20% labeled samples appear to be very similar. As is to be expected, there is an upper limit on the amount of pore former that can be added to increase porosity before collapse of the remaining NiO-YSZ particles take place during sintering.

Figure 4 demonstrates the use of this microdispensing technique for sequential deposition of layers with different pore former amounts in order to grade the structure's porosity. In this case, the bottom two layers of samarium doped ceria (SDC) ink have no added graphite. The two layers following sintering have a thickness of ~30μm (i.e. 15μm thick per layer). The next two layers were printed with SDC ink containing 15 wt% graphite powder.

85 wt% SDC + 15 wt% graphite
2 printed layers
30 μm thick

100 wt% SDC + 0 wt% graphite
2 printed layers
30 μm thick

Figure 4. Graded porous Semarium Doped Ceria (SDC) structure (4 layers printed at 2 different compositions).

ARCHITECTURAL CONTROL VIA DIRECT-WRITE PRINTING

Although graded porosity has been widely considered to be beneficial to SOFC operation, the traditional sponge like porosity that results from most SOFC synthesis techniques can have its drawbacks. Specifically, the high tortuosity associated with sponge-like porosity can:

- Inhibit mass transport

- Increase conductive path length and ohmic losses

- Reduce effectiveness of infiltration techniques

One intriguing synthesis technique that has attracted a great deal of interest in recent years is freeze tape casting[12,13]. With this process, conventional tape casting is done over a chilled platen that freezes the film as the carrier tape is drawn from the doctor blade. As ice crystals form, solid particulate matter is rejected to the boundary regions between the crystals. The frozen tape cast is then freeze dried such that the volume previously occupied by the ice crystals is vacated. Lastly, the green tape is sintered in a conventional furnace. The porosity of the resulting structure is dictated by the size and shape of the ice crystals which can be controlled to a certain extent by ink composition and freezing rate. The formation of ice crystals is governed by direction of heat flux which is highly directional in tape casting. Consequently, freeze tape cast structures tend to have long slender channels rather than round interconnected pores seen with conventional SOFC synthesis techniques. The channel-like architecture has considerable potential for addressing some of the issues mentioned above stemming from high tortuosity.

In this research, the use of direct-write printing has been explored as a potential means of building upon the advantages of freeze tape casting. More specifically, the aims are to demonstrate that it is possible to print channeled electrode architectures in any desired geometry. Having the ability to vary channel geometry is desirable in the sense that architectures better suited to withstand shrinkage stresses and subsequent handling can be designed. Furthermore, direct-write approaches will ultimately allow for spatial control of chemical composition as well.

In order to demonstrate feasibility of direct-write methods for production of channeled electrode structures, it is necessary to stack single-wide lines of ink on top of one another while avoiding slump or collapse of the printed stack while the ink dries. To do this, an NiO-YSZ anode material has been used. The ink was made in the ratio of 12 grams NiO-YSZ, 0.5 grams graphite (4% wt.) and 3ml VEH (Terpineol). The ink was thoroughly mixed in a THINKY mixer for 20 minutes @ 2000 RPM. The ink was then transferred to a 3cc syringe and placed in the nScrypt tool. The powder constituents for the anode disk (substrate) were 8 grams NiO-YSZ powder and 2 grams graphite (TC307). The powder was mixed in a VQN high speed ball mill for 20 minutes. Several drops of PVA solution, created by mixing 5 grams with PVA with 100 ml distilled water was added to the mixed powder. The powder was allowed to dry and was then die pressed at 3 metric tons for 25 seconds.

A series of nested concentric circles having a step-over of 250 μm were printed. This was done to ensure symmetric shrinkage stresses and to produce a patterned structure with greater mechanical strength in all directions for improved ease of handling. A tool path was written for the nScrypt machine to print concentric circles. The initial outer diameter was 6mm (6000 μm) and each successive circle was offset inward by 250 microns (i.e. step-over of 250μm). Lines were printed on the un-sintered die pressed NiO-YSZ anode disk. Concentric rings for the first layer were printed, and the dispensing height was then increased by 50μm. A second layer of concentric rings were then printed on top of the first layer of rings. The dispensing height was then increased by another 50μm and a third layer printed. A total of 6 layers were printed with +50 μm dispensing height adjustments. The process parameters for the above experiment are tabulated in Table II. A 75 μm ceramic nozzle was used for extrusion.

Table II. Process parameter for channel structure with step-over of 250μm

Number of layers	Dispensing Height (mm)	Speed (mm/s)	Pressure (PSI)	Step-over (μm)	Valve Opening (mm)
1	-58.90	0.75	46		2.4
2	-58.85	0.75	46		2.4
3	-58.80	0.75	46	250	2.4
4	-58.75	0.75	46		2.4
5	-58.70	0.75	46		2.4
6	-58.65	0.75	46		2.4

The printed samples were dried under a 250W heat lamp at a distance of 2 feet for one hour. The height of the green (unsintered) structure was ~378μm when measured with a Hirox Microscope (KH-7700). The sample was then sintered at 1400°C for 2 hours at a ramp up rate of 5°C per minute. The height of the sintered sample was ~270 μm (Figure 5). The sample along with the die-pressed disk showed no warping and no delamination.

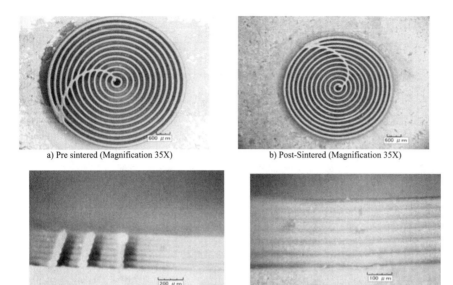

a) Pre sintered (Magnification 35X)
b) Post-Sintered (Magnification 35X)

c) X-section (Post Sintered)
d) Side-view of the line (each layer is visible)

Figure 5. Channeled NiO-YSZ anode architecture

This demonstrates the potential of direct-write printing to produce engineered electrode structures with low tortuosity and balanced shrinkage stresses. Figure 6 (a-d) illustrates how built-up ribs can be produced having any desired overall height with relatively smooth side walls or highly scalloped side walls depending on the choice of process parameters. In Figure 6(e), a stepover distance of 120μm produces ribs that are 60-70μm wide with 20-25μm channels after sintering. Note that the sum of rib and channel width does not equal stepover distance due to the fact that process parameters can be chosen that stretch (narrow) the extruded bead or squash (flatten) it out.

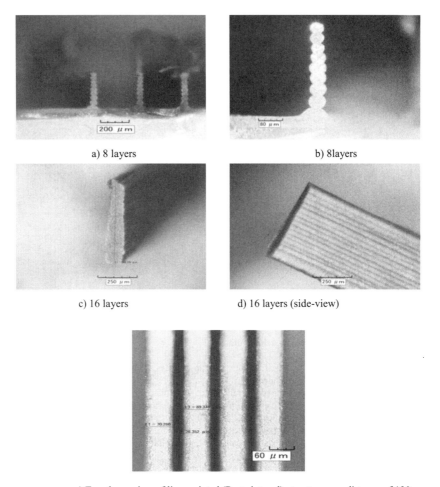

a) 8 layers

b) 8layers

c) 16 layers

d) 16 layers (side-view)

e) Top-down view of lines printed (Post sintered) at a step-over distance of 120μm

Figure 6. Structured Electrodes: 8 layers and 16 layers

In the case where a rib has smooth vertical walls, the surface area and porosity can be analyzed as a function of rib and channel geometry as follows (Figure 7). Assuming that the ribs are approximated by rectangles with height (h), line width (l), channel width (w), length of line (len) and number of lines (*n*):

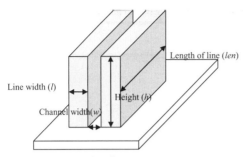

Figure 7. Modeling of smooth vertical walls

$$Porosity\ (assuming\ the\ lines\ are\ fully\ dense) = \frac{Volume\ of\ Void\ Space}{Total\ Volume}$$

$$= \frac{n * w * h * len}{(l + w) \times n \times len \times h}$$

$$Total\ surface\ area\ (all\ 6\ sides) = 2 * n * [(h * len) + (h * l) + (l * len)]$$

The total surface area calculated above includes all six sides of the ribs and assumes that the structure are fully dense, whereas the top/bottom and front/back faces would be covered in a real cell and the structure will be porous. So, the actual total surface area is:

$$Actual\ Total\ surface\ area \geq 2 * n * h * len, where\ n = \frac{Area}{(l + w)}$$

Since the structure is actually porous, the actual total surface area will go up depending on the chosen porosity of the ribs. This arrangement provides channels for the gases to easily flow into. It also provides a better means for solution infiltration techniques commonly used for cathodes.

Using typical values for an anode layer (say a 20mm x 20mm square button cell with h =500 microns thick), Figure 8 shows the effect of channel width *(w* = 20 to 150μm) on actual total surface area for *l* = 50, 75 and 100μm.

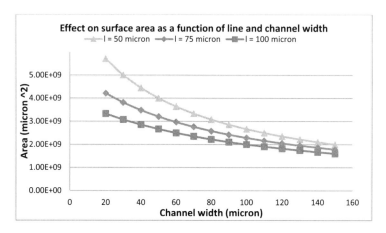

Figure 8. Effect on surface area as a function of channel width

In Figure 6, the ability to produce ribs having fairly smooth or heavily scalloped vertical sidewalls was demonstrated. A rib having scalloped side walls has a greater amount of surface area than ribs with smooth walls. The degree to which process parameters influence surface area in the printed ribs is therefore analyzed. The surface area of the rib is a function of extruded bead diameter and the layer thickness. When a viscous paste is extruded through a round orifice, it will take on a cylindrical shape that may be less than, approximately equal to, or greater than the orifice diameter. If the gap between the orifice and the substrate is less than the bead diameter, then the bead will be "squashed" to a certain extent. By depositing multiple beads on top of each other to create a fin structure, it is possible to create a wall having scallops. The net effect is a wall whose surface area is greater than that of a perfectly vertical wall.

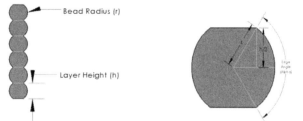

With a perfectly vertical wall, the edge length equals h. With a scalloped wall as shown above, the scalloped edge length can be calculated as a function of the bead radius (r) and layer height (h). The angle that is swept by the scalloped edge (α) is determined as follows:

$$\sin\left(\frac{\alpha}{2}\right) = \frac{h/2}{r}$$

$$\alpha = 2\sin^{-1}\left(\frac{h}{2r}\right)$$

Since the circumference of a circle having radius r is πd, the length of the arc swept through an angle α on that circle is determined as follows:

$$Edge\ Length\ (one\ side) = \frac{\alpha}{360}\pi d = \frac{\sin^{-1}\left(\frac{h}{2r}\right)\pi d}{180}$$

Table III computes the scallop edge length as a function of layer height and radius (assuming that a radius = 50 μm). If the scallop edge length is compared with a vertical wall edge length of h, then the percent increase in edge length grows rapidly as the layer height increases relative to a given bead radius. Since surface area is simply edge length multiplied by the length of the rib, the increase in overall electrode surface area scales directly with the increase in edge length. This increase in surface area is also achieved with very little influence on the tortuosity of the electrode, but could have a negative impact on the mechanical properties of the structure.

Table III. Scalloped Edge length as a function of layer height and radius

Layer Height (h) (μm)	Radius (r) (μm)	Scallop Edge Length (μm)	Percent Increase In Edge Length
50	50	52.4	4.7%
60	50	64.4	7.3%
70	50	77.5	10.8%
80	50	92.7	15.9%
90	50	112.0	24.4%
100	50	157.1	57.1%

By way of example, the 8 layer structure shown in Figure 6 has an average diameter of 46.5 μm ± 5.5 μm and an average layer height of 36.4 μm ± 4.8 μm. The ink was extruded though a 75 μm nozzle. The ratio of the bead radius of the extruded layer to the nozzle size is 0.62 (= 46.5/75). Using the formula above for calculating the edge length, an increase of 14.8% surface area was calculated for scalloped line when compared to vertical wall.

$$Vertical\ height = 36.4\ \mu m$$

$$Scallop\ Edge\ Length = \frac{\sin^{-1}\left(\frac{h}{2r}\right)\pi d}{180} = \frac{\sin^{-1}\left(\frac{36.4}{46.5}\right)\pi * 46.5}{180} = \frac{51.5 * \pi * 46.5}{180} = 41.8\ \mu m$$

$$Increase\ in\ surface\ area\ (vs.\ vertical\ wall) = \frac{41.8 - 36.4}{36.4} = 14.8\%$$

SUMMARY AND CONCLUSIONS

This paper has presented preliminary results associated with the use of a microextrusion-based direct write printing process to produce heterogeneous SOFC materials. This initial phase of research has emphasized the porosity of the anode and cathode regions. The technique is quite well suited for synthesis of porous layers having graded porosity. Although it has not been presented here, a multi-material microextrusion tool can be used to increase or decrease the amount of graphite pore former that is blended into the SOFC ink on the fly.

While graded porosity is useful, efforts are currently underway to explore novel channel architectures that could provide added functionality. Cells having concentric circles have been fabricated which provide more balanced shrinkage stresses and higher handling strength when compared with linear channel architectures. The next step will be to evaluate relationships between rib and channel size and porosity and the resulting electrochemical performance. In addition to the work presented here involving graded porosity and channeled electrodes, work is also underway on graded material compositions.

ACKNOWLEDGEMENTS

The authors gratefully acknowledge the financial support of the HeteroFoam EFRC, an Energy Frontier Research Center funded by the U.S. Department of Energy, Office of Basic Energy Sciences under Award Number DE-SC0001061.

REFERENCES

[1] Hae-Gu Park, Hwan Moon, Sung-Chul Park, Jong-Jin Lee, Daeil Yoon, Sang-Hoon Hyun, Do-Heyoung Kim, "Performance improvement of anode-supported electrolytes for planar solid oxide fuel cells via a tape-casting/lamination/co-firing technique", Journal of Power Sources, Volume 195, Issue 9, 1 May 2010, Pages 2463-2469.

[2] D. Rotureau, J.-P. Viricelle, C. Pijolat, N. Caillol, M. Pijolat, "Development of a planar SOFC device using screen-printing technology," Journal of the European Ceramic Society, Volume 25, Issue 12, Elecroceramics IX, 2005, Pages 2633-2636.

[3] J. Ding, J. Liu, "An anode-supported solid oxide fuel cell with spray-coated yttria-stabilized zirconia (YSZ) electrolyte film", Solid State Ionics 179 (2008), 1246–1249.

[4] Rob Hui, Zhenwei Wang, Olivera Kesler, Lars Rose, Jasna Jankovic, Sing Yick, Radenka Maric, Dave Ghosh, "Thermal plasma spraying for SOFCs: Applications, potential advantages, and challenges", Journal of Power Sources, Volume 170, Issue 2, 10 July 2007, Pages 308-323.

[5] Wang Z., Weng W., Chen K., Shen G., Du P., Han G, "Preparation and performance of nanostructured porous thin cathode for low-temperature solid oxide fuel cells by spin-coating method." (2008) Journal of Power Sources, 175 (1), pp. 430-435.

[6] K. L. Reifsnider, F. Rabbi, R. Raihan, Q. Liu, P. Majumdar, Y. Du and J.M. Adkins " HeteroFoam: New Concepts and Tools for Heterogeneous Functional Material Design"

[7] B. Li, K.H. Church, P.A. Clark, "Robust Direct-Write Dispensing Tool and Solutions for Micro/Meso-Scale Manufacturing and Packaging," ASME Conf. Proc. 2007, 715 (2007).

[8] Xu, X., Xia, C., Xiao, G., and Peng, D., 2005, "Fabrication and Performance of Functionally Graded Cathodes for It-Sofcs Based on Doped Ceria Electrolytes," Solid State Ionics, 176(17-18), pp. 1513-1520.

[9] Hart, N. T., Brandon, N. P., Day, M. J., and Lapena-Rey, N., 2002, "Functionally Graded Composite Cathodes for Solid Oxide Fuel Cells," Journal of Power Sources, 106(1-2), pp. 42-50.

[10] Ni, M., Leung, M. K. H., and Leung, D. Y. C., 2007, "Micro-Scale Modeling of Solid Oxide Fuel Cells with Micro-Structurally Graded Electrodes," Journal of Power Sources, 168(2), pp. 369-378.

[11] Ni, M., Leung, M. K. H., and Leung, D. Y. C., 2007, "Micro-Scale Modeling of a Functionally Graded Ni-Ysz Anode," Chemical Engineering and Technology, 30(5), pp. 587-592.

[12] Stephen W. Sofie, "Fabrication of Functionally Graded and Aligned Pores Via Tape Cast Processing", J. Am. Ceram. Soc., 90 [7] 2024–2031 (2007).

[13] F. Dogan and S.W. Sofie, "Microstructural Control of Complex-Shaped Ceramics Processed by Freeze Casting." Ceram.Forum Int./Ber.DKG. 79 [5], pp E35-E38 (2002).

MANGANESE COBALT SPINEL OXIDE BASED COATINGS FOR SOFC INTERCONNECTS

Jeffrey W. Fergus, Yingjia Liu and Yu Zhao
Auburn University
Auburn, AL, USA

ABSTRACT
    Ferritic stainless steels have good oxidation resistance and an appropriate coefficient of thermal expansion for use as SOFC interconnects. The chromia scale is desirable because it has a higher electrical conductivity than other common protective oxide scales, such as alumina and silica. However, chromia volatilization can lead to cathode poisoning, so ceramic coatings are applied to minimize this degradation. Manganese cobalt spinel oxide coatings have been shown to reduce chromium volatilization and maintain a low electrical resistance. However, a reaction layer does form and could lead to performance degradation for long exposures. In this paper results on the properties of this reaction layer are presented.

INTRODUCTION
    Solid oxide fuel cells (SOFCs) can operate with a variety of fuels, which expands the range of potential applications[1]. This fuel flexibility is a result of the high operating temperatures, which increase electrode reaction rates, but also increase the rates of undesired reactions and thus create materials challenges[2]. One example is the interconnect material, which is exposed to both air and fuel atmospheres[3-5]. Chromia-forming alloys are generally used because the scale formed during oxidation is protective but also has reasonable electrical conductivity (as compared to alumina or silica). However, chromia can be oxidized in air to form volatile species that are reduced to form Cr_2O_3 on the cathode and can degrade cell performance[6]. Although the amount of volatilization can be reduced by alloying, long operation times require further reduction in the amount of chromium volatilization, which can be accomplished with ceramic coatings[7]. One promising coating material system is the spinel $(Mn,Co)_3O_4$[8-11], which has been shown to reduce chromium volatilization[12,13]. During long exposure, interaction between the coating and the scale formed on the alloy can affect the coating performance. We have previously shown that the reaction between $(Mn,Co)_3O_4$ and Cr_2O_3 results in a two-layer reaction zone[14]. Both zones form the spinel structure, but grow by different mechanisms. The phase with the higher chromium content grows by the diffusion of manganese and cobalt from the coating toward the Cr_2O_3, while another layer grows as chromium diffusions into the coating. The purpose of this paper is to report results from characterization of the transport properties of some $(Mn,Co,Cr)_3O_4$ phases that are relevant to understanding the effect of coating-scale interactions on the long-term performance of the coating.

EXPERIMENTAL
    Pellets of $(Mn,Co,Cr)_3O_4$ were prepared by solid-state synthesis from MnO (99%), Co_3O_4 (99.83%) and Cr_2O_3 (99%) powders. The powders were mixed with deionized water, ball milled for 48 hours, dried overnight, pressed into pellets and then sintered in air at 1200°C for 24 hours.
    The conductivities of coating materials were determined using 4-point dc conductivity measurements on rectangular bars (13-14 mm x 5-6 mm x 1.5-2 mm) that were sectioned from sintered pellets (25 mm diameter, 1.5-2 mm thickness).

RESULTS AND DISCUSSION

Figure 1 shows that the conductivity of $(Mn,Co)_3O_4$ decreases with the addition of chromium. With small amounts of chromium (up to about $x = 0.5$ in $Mn_{1.5-0.5x}Co_{1.5-0.5x}Cr_xO_4$) the decrease is relatively small and the conductivity is still in the same range as the conductivity of the original coating materials. However, with higher chromium contents, the conductivity is orders of magnitude lower than the conductivity of the coating. These results are summarized in Figure 2, which shows that there is a significant increase in activation energy for conduction for x greater than 1.

The results in Figures 1 and 2 suggest that the electrical resistance between the cathode and the interconnect, and thus the area specific resistance (ASR) of the cell, will likely be dominated by the oxides with high chromium content – the Cr_2O_3 scale or the high-chromium $(Mn,Co,Cr)_3O_4$ spinel layer. However, the conductivity also depends on the cobalt and manganese contents. Figure 3 shows the effect of Mn/Co ratio on the conductivity of $(Mn,Co,Cr)_3O_4$ spinel phases with two different chromium contents. In both cases, the conductivity increases with increasing cobalt content. This is consistent with a recent report by Stevenson et al.[16], which showed that the ASR of $(Mn,Co)_3O_4$ coatings decreased with increasing cobalt content.

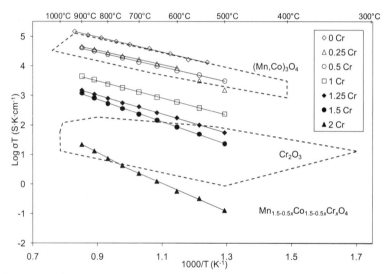

Figure 1. Conductivity of $Mn_{1.5-0.5x}Co_{1.5-0.5x}Cr_xO_4$ in air. Conductivity ranges for $(Mn,Co)_3O_4$ and Cr_2O_3 summarized from several sources[15].

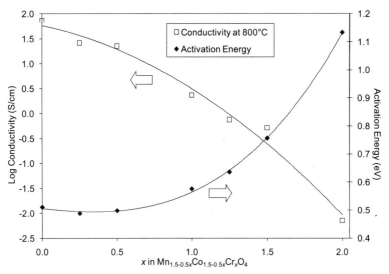

Figure 2. Conductivity and activation energy of $Mn_{1.5-0.5x}Co_{1.5-0.5x}Cr_xO_4$ in air.

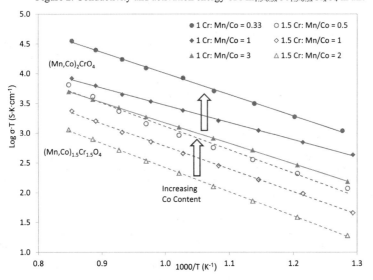

Figure 3. Conductivity of $Mn_{1.5-0.5x}Co_{1.5-0.5x}Cr_xO_4$ in air.

Stevenson *et al.*[16] also reported that the ASR of $(Mn,Co)_3O_4$-coated 441 stainless steel reached a constant value with time for exposures of up to 18,000 hours. Even with parabolic growth a continual increase in scale thickness, and thus ASR, is expected. However, a constant ASR could be explained by the reaction mechanism mentioned above. In particular, the high-chromium layers are growing at the expense of the alloy, but at the same time are being consumed as chromium diffuses into the coating. In such a situation, it is possible that the thickness of the high-chromium layers, and thus the associated electrical resistance, could remain constant with time.

CONCLUSIONS

The conductivity of $(Mn,Co,Cr)_3O_4$ decreases with increasing chromium and manganese contents. The high chromium-content spinel phase and the chromia scale have the lowest electrical conductivity and thus will likely dominate the area specific resistance during long-term SOFC operation.

ACKNOWLEDGMENTS

Financial support from the Department of Energy through the Building EPSCoR-State/National Laboratory Partnerships Program (Timothy Fitzsimmons, Program Officer) and the National Energy Technology Laboratory (NETL) (Briggs White, Program Officer) is gratefully acknowledged.

REFERENCES

[1] E.D. Wachsman and S.C. Singhal, Solid Oxide Fuel Cell Commmercialization, Research and Challenges, *ECS Interface*, **18**[3], 38-43 (2009).

[2] A.J. Jacobson, Materials for Solid Oxide Fuel Cells, *Chem. Mater.*, **22**, 660-674 (2010).

[3] W.J. Quadakkers, J. Piron-Abellan, V. Shemet and L. Singheiser, Metallic Interconnectors for Solid Oxide Fuel Cells – A Review, *Materials at High Temperatures*, **20**, 115-127 (2003)..

[4] J.W. Fergus, Metallic Interconnects for Solid Oxide Fuel Cells, *Mater. Sci. Eng. A*, **397**, 271-283 (2005).

[5] Z. Yang, K.S. Weil, D.M. Paxton and J.W. Stevenson, Selection and Evaluation of Heat-Resistant Alloys for SOFC Interconnect Applications, *J. Electrochem. Soc.*, **150**, A1188-A1201 (2003).

[6] J.W. Fergus, Effect of Cathode and Electrolyte Transport Properties on Chromium Poisoning in Solid Oxide Fuel Cells, *Int. J. Hydrogen Energy*, **32**[16], 3664-3671 (2007).

[7] N. Shaigan, W. Qu, D.G. Ivey and W. Chen, A Review of Recent Progress in Coatings, Surface Modifications and Alloy Developments for Solid Oxide Fuel Cell Ferritic Stainless Steel Interconnects, *J. Power Sources*, **195**, 1529-1542 (2010).

[8] Y. Larring and T. Norby, Spinel and Perovskite Functional Layers between Plansee Metallic Interconnect (Cr-5 wt% Fe-1 wt% Y_2O_3) and Ceramic $(La_{0.85}Sr_{0.15})_{0.91}MnO_3$ Cathode Materials for Solid Oxide Fuel Cells, *J. Electrochem. Soc.* **147**, 3251-3256, (2000).

[9] Z. Yang, G.-G. Xia, G.D. Maupin and J.W. Stevenson, Conductive Protection Layers on Oxidation Resistance Alloys for SOFC Interconnect Applications, *Surf. Coating Tech.*, **201**, 4476-4483 (2006).

[10] M.R. Bateni, P. Wei, X. Deng and A. Petric, Spinel Coatings for UNS 430 Stainless Steel Interconnects, *Surf. Coating Tech.* **201**, 4677-4684 (2007).

[11]M. J. Garcia-Vargas, M. Zahid, F. Tietz and A. Aslanides, Use of SOFC Metallic Interconnect Coated with Spinel Protective Layers using the APS Technology, *ECS Trans.* **7**, 2399-2405 (2007).

[12]H. Kurokawa, C.P. Jacobson, L.C. DeJonghe and S.J. Visco, Chromium Vaporization of Bare and of Coated Iron-Chromium Alloys at 1073 K, *Solid State Ionics*, **178**, 287-296 (2007).

[13]C. Collins, J. Lucas, T.L. Buchanan, M. Kopczyk, A. Kayani, P.E. Gannon, M.C. Deibert, R.J. Smith, D.-S. Choi and V.I. Gorokhovsky, Chromium Volatility of Coated and Uncoated Steel Interconnects for SOFCs, *Surf. Coating Tech.* **201**, 4467-4470 (2006).

[14]K. Wang, Y. Liu and J.W. Fergus, Interactions between SOFC Interconnect Coating Materials and Chromia, *J. Am. Ceram. Soc.* in press. doi: 10.1111/j.1551-2916.2011.04749.x

[15]J.W. Fergus, K. Wang and Y. Liu, "Transition Metal Spinel Oxide Coatings for Reducing Chromium Poisoning in SOFCs," *Electrochem. Trans.* **33**[40] (2011) 77-84.

[16]J.W. Stevenson, G.G. Xia, J.P. Choi, Y.S Chou, E.C. Thomsen, K.J. Yoon, R.C. Scott, X. Li and Z. Ĭie, Development of SOFC Interconnects and Coatings, *Proc. 12th Ann. SECA Workshop*, (DOE, 2011).

CO$_2$ CONVERSION INTO C/CO USING ODF ELECTRODES WITH SOEC

Bruce Kang[1], Huang Guo[1], Gulfam Iqbal[1]
Ayyakkanna Manivannan[2]
[1] Mechanical and Aerospace Engineering Department, West Virginia University
[2] Energy System Dynamics Division, National Energy Technology Laboratory

ABSTRACT

Oxygen-Deficient Ferrite (ODF) electrodes are integrated with the YSZ electrolyte to decompose carbon dioxide (CO$_2$) into solid carbon (C) or carbon monoxide (CO) in a continuous process. The cells are tested in a NexTech Probostate^{TM} apparatus and the performance is evaluated via EIS/potentiostate and gas chromatography (GC). In preliminary tests, high percentages of CO and O$_2$ was detected at the cathode and anode sides flue gases respectively, when CO$_2$ was fed to the cathode side under a small potential bias applied across the electrode. Depending on the applied potential, the system is capable of decomposing CO$_2$ into C or CO. Through the in-situ EIS and flue gas analyses, and post-test material characterization techniques, the capability and efficiency of process are currently being evaluated.

INTRODUCTION

The attenuation of CO$_2$ concentration, emitted mainly from fossil fuel power plants, has been an important ecological issue associated with the global warming. The state-of-the-art technology for CO$_2$ capturing involves two steps: (i) chemical absorption of CO$_2$ using monoethanolamine (MEA) and (ii) CO$_2$ collection by heating up the MEA to release the captured CO$_2$ to a storage unit. This process consumes a significant portion of the power plant energy output and the captured CO$_2$ must be sequestrated to a permanent place which is another energy-consuming process [1-6]. A preferable approach would be to decompose CO$_2$ into solid carbon/CO and O$_2$, or co-electrolysis CO$_2$ with H$_2$O to generate syngas (H$_2$+CO) and O$_2$. Currently, two main approaches have been actively under development to decompose CO$_2$/H$_2$O into H$_2$/CO; (i) CO$_2$/H$_2$O-splitting thermochemical cycle with metal oxides (e.g. Fe$_3$O$_4$) using highly concentrated solar heat [7-10] and (ii) co-electrolysis of CO$_2$/H$_2$O in the solid oxide electrolyser cell (i.e. reverse SOFC) [11-17]. The first approach requires high temperature thermal reduction (>1350°C) on the metal oxides and the repeated thermo-chemical cycles significantly limit lifetime of the metal oxides due to melting, sintering and spoliation [10]. In the second approach, high electric energy is needed to activate the electrolysis of CO$_2$/H$_2$O of the reverse SOFC approach

The reduced metal oxides are known to have high (near 100%) CO$_2$/H$_2$O-splitting capability [18-20]. Metal oxides have also been used to decompose CO$_2$ at lower temperature [18-20] but this process is not continuous and requires a reducing agent (e.g. H$_2$) intermittently to reactivate the metal oxides. In this research, we demonstrated an integration of oxygen deficient ferrite (ODF) based electrodes with Solid Oxide Electrolyser Cell (SOEC) to decompose CO$_2$/H$_2$O into C/CO/H$_2$ and O$_2$ that can also be used to decompose NO$_x$ into N$_2$ and O$_2$. Initial tests showed promising results for CO$_2$ decomposition by electrolysis.

TEST METHODOLOGY

The principle of ODF reactivity is shown in Figure 1. The ODF has been demonstrated to have high efficiency to decompose CO$_2$ to atomic carbon [18-20]. ODF (M$_x$Fe$_{3-x}$O$_{4-\delta}$) is formed by the reducing the spinal ferrites (M$_x$Fe$_{3-x}$O$_4$) with hydrogen gas (H$_2$) as shown in Eq. (1). Here M represents a bivalent metal ion such as Fe(II), Co(II), Mn(II), Ni(II), Cu (II) and so on; the oxygen deficiency (δ expresses the degree of reduction. The ODF then decomposed CO$_2$ to carbon as shown in Eq. (2). During the CO$_2$ decomposition, carbon is deposited on the ODF surface and oxygen transferred in the form of O^{2-} to be incorporated into the vacant lattice sites of ODF. The deposited carbon powder can be separated by mechanical or chemical processes, or can be converted into methane or syngas (e.g. Eq. (3)).

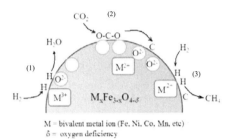

Figure 1. Principle of ODF reactivity (redrawn from [20])

$$H_2 + O^{2-} + 2Fe^{3+} \rightarrow H_2O + Vo + 2Fe^{2+} \qquad (1)$$

$$CO_2 + 2Vo + 4Fe^{2+} \rightarrow C + 2O^{2-} + 4Fe^{3+} \qquad (2)$$

$$C + 2H_2 \rightarrow CH_4 \qquad (3)$$

Based on the working principles of ODF and SOEC technology, we designed an integrated CO_2/H_2O decomposition/fuel cell energy conversion system, as shown in Figure 2. Analogous to the fuel cell structural construction, the unit cell consists of dense Yttria Stabilized Zirconia (YSZ) electrolyte, as ionic-oxygen conductor, and nano-ODF (e.g. nickel ferrite) particles with YSZ powders as electrodes for both cathode and/or anode. Corresponding to the reactions shown in Eqs. (1)-(3), when CO_2 is fed to the cathode side and H_2 is fed to the anode side, the electrode reactions can be described as following:

Anode reaction: $\qquad\qquad 2O^{2-} \rightarrow O_2 + 4e^- \qquad\qquad\qquad$ (4)

Cathode reaction: $\qquad\qquad 2CO_2 + 4e^- \rightarrow 2CO + 2O^{2-} \qquad$ (5)

$\qquad\qquad$ or $\quad CO_2 + 4e^- \rightarrow C + 2O^{2-} \qquad\qquad$ (6)

The overall reaction: $\qquad\qquad 2CO_2 \rightarrow 2CO + O_2 \qquad\qquad\qquad$ (7)

$\qquad\qquad$ or $\quad CO_2 \rightarrow C + O_2 \qquad\qquad\qquad\qquad$ (8)

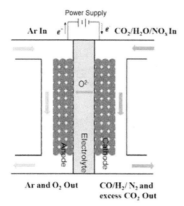

Figure 2. Schematic of $H_2O/CO_2/NO_x$ decomposition into $H_2/CO/N_2$ and O_2

It should be noted that the CO_2 decomposition reactivity is due to the oxygen vacancies on the ODF surface. It is reported that Ni-ferrite loses 24 wt% of the oxygen in structure in the reduction reaction up to 800°C [21]. Without the integrating fuel cell design, the oxygen vacancy will be occupied by the oxygen in the CO_2 and ODF gradually lose its catalytic activity. In our system, due to the electric potential, the oxygen will be transported to the anode side through YSZ electrolyte in the form of O^{2-} and react with H_2. This process will keep the oxygen vacancies on the ODF surface and extend its catalytic activity period. A combination of CO_2, H_2O, and/or H_2, can be supplied to the cathode side instead of pure CO_2 for CO_2 decomposition. The byproduct in this process, syngas $(CO+H_2)$, and/or CH_4, also have high economic value. Compared to other metal electrodes in conventional SOFCs, the ODF electrode material has demonstrated higher catalytic activity under lower temperature and can spontaneously decompose CO_2 into carbon at temperature as low as 290°C even without loading voltage. Thermodynamic calculations show that the decomposing reaction based on ODF is slightly exothermic [7-9], e.g. $\Delta H°_{298K}$ = -33.6 kJ/mol for $3FeO + H_2O \rightarrow Fe_3O_4 + H_2$. In this method, the contribution of loading voltage and electrolyte materials (e.g. YSZ powder and substrate) is for ionic oxygen transfer and to maintain the ODF in reduced condition. Therefore, the proposed technique will significantly reduce the electric energy requirement as compared to conventional SOEC techniques.

EXPERIMENTAL ARRANGEMENT AND PRELIMINARY RESULTS
The catalytic and electrocatalytic measurements were obtained in the ODF/SOEC system shown in Figure 3. The test setup consists of a button cell mounted inside a NexTech Probostat™ apparatus integrated with EIS/Potentiostat and Gas Chromatography (GC).

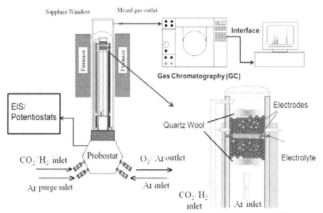

Figure 3. ODF/SOEC system inside the NexTech Probostat^{TM} test apparatus

The button cell was constructed using a 0.3mm thick YSZ (8 mole% yttria) electrolyte that was sealed inside the NexTech Probostat^{TM} using AREMCO-516 high temperature cement. NiFe$_2$O$_4$ Powder (50 nm) was filled in both sides of YSZ electrolyte that serve as anode and cathode. Ag mesh was embedded in the metal powered as current collector on both sides. The electrodes were supported by the quartz wool to keep them in place. Ag current cables and voltage taps were spot welded on each current collector mesh. Alicat^{TM} mass flow controllers (MFCs) were used to control fuel/air flow rates, pressure and fuel compositions.

The button cell was heated from room temperature to 750 °C at a rate of 1 °C/min. During this period, the anode and cathode were supplied with 50 sccm of Ar. After that, 50 sccm H$_2$ was provided to the cathode to reduce NiFe$_2$O$_4$ Powder into ODF at 750 °C. Once the reduction of electrodes was completed, the Ar flow rate was increased to 100 sccm on anode side, while the cathode was exposed to 30 sccm CO$_2$ with 10% H$_2$. The experiment investigation was carried out at 750 °C and the cell electrochemical performances were measured using Reference 300 Potentiostat/Galvanostat/ZRA (Gamry Instruments, Warminster, PA), and exhaust gases were analyzed via Gas Chromatography (GC).

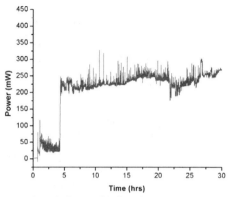

Figure 4. Potentiostatic Curve of Cell Performance under 800 mV potential bias at 750 °C

Table 1 Gas Chromatography Analysis

Compound	Description									
	After decomposition for 6 hrs		After decomposition for 150 hrs		After decomposition for 461 hrs		After decomposition for 531 hrs		After decomposition after 672 hrs	
	Cathode Side (%)	Anode Side (%)	Cathode Side (%)	Anode Side (%)	Cathode Side (%)	Anode Side (%)	Cathode Side (%)	Anode Side (%)	Cathode Side (%)	Anode Side (%)
CO_2	44.72	3.04	49.48	5.06	47.23	3.41	44.20	5.05	50.99	3.41
CO	*I D*	*I D*	*I D*	*I D*	*1.16*	*I D*	*0.52*	*0.14*	*2.17*	*I D*
O₂	*13.35*	*79.32*	*13.16*	*94.16*	*14.05*	*56.08*	*13.94*	*84.48*	*14.04*	*66.14*
H_2	ND	ND	ND	ND	ND	ND	ND	ND	ND	ND
Ar	8.17	ND	8.45	ND	4.84	8.94	3.79	9.55	NA	NA
N_2	33.76	17.64	28.91	0.78	32.72	31.58	37.34	0.77	32.36	29.98

Figure 4 illustrates the cell performance history under an 800 mV potential bias. In this experiment, CO_2 and H_2 were supplied to the cathode and anode side respectively and a small potential was applied across the electrodes. In the flue gas CO was detected continuously at the cathode side that is shown in Table 1. The tests show promising results as a high percentage of CO_2 decomposition into CO and O_2 was achieve, while the electrolyser unit run continuously for more than 820 hrs under low operating voltage before the experiment was arbitrarily terminated due to the contacts failure. It also shows high potential of further CO electrolysis to produce solid carbon. This test confirmed the feasibility of CO_2/H_2O electrolysis via ODF electrodes for a long-term continuous and highly stable process. The low current density may be due to the low fuel gas flow rate and high contact resistant, which are being improved in the on-going tests.

CONCLUSIONS

In this research, oxygen-deficient ferrite (ODF) electrodes are integrated with the high temperature fuel cell technology for CO_2 decomposition and energy conversion. Initial tests demonstrated its capability of electrochemically reducing CO_2 into CO/C with high efficiency. High percentages of CO and O_2 was detected at the cathode and anode sides flue gases respectively, when CO_2 was fed to the cathode side under a small potential bias applied across the electrode. Further experimental work and results will be reported in a future paper.

REFERENCES

[1]Shinichirou Morimoto, Kotarou Taki, and Tadashi Maruyam, Current review of CO_2 separation and recovery technologies. In 4th Workshop of International Test Network for CO_2 capture, Kyoto, Japan, October 2002. IEA Greenhouse Gas R&D Programme.

[2]John Marion, Nsakala ya Nsakala, Carl Bozzuto, Gregory Liljedahl, Mark Palkes, David Vogel, J. C. Gupta, Manoj Guha, Howard Johnson, and Sean Plasynski, Engineering feasibility of CO_2 capture on an existing US coal-fired power plant. In 26th International Conference on Coal Utilization & Fuel Systems, Clearwater, Florida, March 2001.

[3]David J. Singh. Simulation of CO2 capture strategies for an existing coal fired power plant - MEA scrubbing versus O_2/CO_2 recycle combustion. Master's thesis, University of Waterloo, 2001.

[4]Umberto Desideri and Alberto Paolucci, Performance modeling of a carbon dioxide removal system for power plants. Energy Conversion and Management, 40, 1899-1915 (1999).

[5]Carl Mariz, LarryWard, Garfiled Ganong, and Rob Hargrave, Cost of CO2 recovery and transmission for EOR from boiler stack gas. In Pierce Riemer and Alexander Wokaun, editors, Greenhouse Gas Control Technologies: Proceedings of the 4th International Conference on Greenhouse Gas Control Technologies. Elsevier Science Ltd., April 1999.

[6]A. Chakma, A. K. Mehrotra, and B. Nielsen. Comparison of chemical solvents for mitigating CO2 emissions from coal-fired power plants, *Heat Recovery Systems and CHP,* 15(2), 231–240 (1995).

[7]T. Kodama and N. Gokon, Thermochemical Cycles for high-Temperature Solar hydrogen Production, *Chem. Rev.,* 107, 4048-4077 (2007).

[8]P. G. Loutzenhiser, A. Stamatiou, W. Villasmil, A. Meier, and A. Steinfeld, Concentrated Solar Energy for Thermochemcially Producing Liquid Fuels from CO_2 and H_2O, *JOM,* 63, 32-34 (2011).

[9]A. Stamatiou, P. G. Loutzenhiser, and A. Steinfeld, Solar Syngas Prodcution from H_2O and CO_2 via Two-Step Thermochemical Cycles Based on Zn/ZnO and FeO/Fe₃O₄ Redox Reactions: Kinetic Analysis, *Energy Fuels,* 24, 2716-2722 (2010).

[10]A. Ambrosini, E. N. Coker, M. A. Rodriguez, S. Livers, L. R. Evans, J. E. Miller, and E. B. Stechel, Synthesis and Characterization of Ferrite Materials for Thermochemical CO_2 Splitting Using Concentrated Solar Energy, Advances in CO₂ Conversion and Utilization, Chapter 1, 1-13, American Chemical Society 2010

[11]F. Bidrawn, G. Kim, G. Corre, *et al.,* Efficient Reduction of CO_2 in a Solid Oxide Electrolyzer, *Electrochmical and Solid-State Letters,* 11(9), B167-B170 (2008).

[12]R. Bredesen, K. Jordal, Olav Bolland, High-temperature membranes in power generation with CO_2 capture, Chemical Engineering and Processing, 43, 1129-1158 (2004).

[13]M. R. Haines, W. K. Heidug, K. J. Li, J. B. Moore, Progress with the development of a CO2 capturing solid oxide fuel cell, *Journal of Power Sources,* 106, 377-380 (2002).

[14]A. Amorelli, M. B. Wilkinson, P. Bedont, *et al,* An experimental investigation into the use of molten carbonate fuel cells to capture CO_2 from gas turbine exhaust gases. *Energy,* 29, 1279-1284 (2004).

[15]C. M. soots, High-Temperature Co-Electrolysis of H_2O and CO_2 for Syngas Production, 2006 Fuel Cell Seminar, INL/CON-06-11719.

[16]B. Yildiz, K. J. Hohnholt and M. S. Kazimi, Hydrogen Production using High-temperature Steam Electrolysis Supported by Advanced gas Reactors with Supercritical CO_2 Cycles, *Juclear Technology,* 155, 1-21 (2006).

[17]Q. Fu, C. Mabilat, M. Zahid, A. Brisse, and L. Gautier, Syngas production via high-temperature steam/CO_2 co-electrolysis: an economic assessment, *Energy & Environmental Science,* 3, 1382-1397 (2010).

[18]Yutaka Tamaura, Masahiro Tabata, Complete reduction of carbon dioxide to carbon using cation-excess magnetite, Nature, 346, 255-256 (1990).

[19]T. Kodama, T. Sano, T. Yoshida, M. Tsuji and Y. Tamaura, CO_2 Decomposition to carbon with ferrite-derived metallic phase at 300°C, Carbon 33 (10), 1443-1447 (1995).

[20]S. Komarneni, M. Tsuji, Y. Wada, et al., Nanophase ferrites for CO2 greenhouse gas decomposition, J. Mater. Chem, 7(12), 2339-2340 (1997).

[21]H.C.Shin, J. H. Oh, B. C. Choi, and S. C. Choi, Design of an energy conversion system with decomposition of H_2O and CO_2 using ferrites, Phys. Stat. Sol. (c)1, No.12, 3748-3753 (2004).

HETEROFOAM: NEW CONCEPTS AND TOOLS FOR HETEROGENEOUS FUNCTIONAL MATERIAL DESIGN

K. L. Reifsnider, F. Rabbi, R. Raihan, Q. Liu, P. Majumdar, Y. Du and J.M. Adkins
Department of Mechanical Engineering, University South Carolina, Columbia, SC 29209,
Reifsnider@engr.sc.edu

ABSTRACT

The present paper will discuss the establishment of basic understanding, generation of physical data, and formulation of science-based theory and computational methods to provide a fundamental foundation for the conceptual design, simulation and fabrication of nano-structured heterogeneous materials for energy systems. The principle objective of this work at the fundamental level is the creation of new revolutionary materials. Our approach is to focus on the associated control science, i.e., the creation of nano-synthesis concepts and processes that control the nano-structural configurations and interfaces (function and geometry) of the active phases. That is the objective of the "HeteroFoaM Center," a DOE Energy Frontiers Research Center that focuses on these topics. The paper will discuss computational and analysis methods from global multiphysics to local electronic scale, new understandings of nano-composite concepts, new supporting experimental tools, and new concepts and methodologies that directly connect theory to "materials by design" strategies that we are using to bring this new horizon within our grasp. New experimental room-temperature, non-invasive interrogation methods for establishing internal integrity and for following material state changes in solid oxide fuel cells (SOFCs) as a function of operation history will also be introduced and discussed.

HETEROFOAM

Heterogeneous functional materials are pervasive in energy systems. Some examples appear in Figure 1 below.

Figure 1: Heterogeneous functional materials for (a) hydrogen storage, (b) batteries, (c) solid oxide fuel cells, and (d) polymer electrolyte fuel cells.

These are *material systems* that consist of multiple materials combined at multiple scales (from nano- to macro-) that actively interact during their functional history in a manner that controls their collective performance as a system at the global level. Examples include composite mixed-conductors, nano- or micro-structured heterogeneous materials, mechanical alloys, nano-structured interfaces and heterostructures, and many other combinations that typically serve as the heart of engineering devices such as fuel cells, electrolyzers, electrodes, photovoltaics, combustion devices, fuel processing devices, and functional membranes and coatings. The functional behavior of these materials occurs at multiple scales of time and length and is controlled by multiphysics interactions of the material phases.

HeteroFoam materials are related to heterostructures and heterojunctions which are the heart of many of the semiconductor devices that have changed our society so dramatically.[1] In heterostructures, it is typical, in semiconducting materials for example, to create a material boundary by epitaxially growing one material on the crystalline surface of another, selected in such a way as to create specific alignments of valance and conduction bands in the two materials. Field effect transitors, light-emitting diodes and lasers are made possible by such heterogeneous material creations. Heterostructures typically involve precise electrical design, and very exact mechanical matching of the crystalline structures across the junction. However, HeteroFoaMs are heterogeneous functional materials that, unlike the heterostructures common to electronic materials, involve the co-dependent management, conversion, and transfer of mass, heat and electricity. For a fuel cell, for example, there must be material porosity to transport fuel and air, electrochemical surfaces to oxidize and reduce the reactants, electrically conductive materials (typically both electric and ionic conductors) to create an electric circuit, and material conduits to manage heat flow. However, we can learn our first essential lesson from this comparison. The unique and definitive properties and functionality of heterostructures and HeteroFoaMs are determined by the interaction of the materials across their boundaries (which may be disordered). And those global properties and functionalities may not be part of the character of any of the individual constituents (including reactants). *In order to create such material systems, the multiphysics (physics, chemistry ,and mechanics) of the interactions must be understood and correctly modeled.*

A second basic concept associated with HeteroFoaMs is illustrated by considering the fact that they are also structural composites. Batteries, fuel cells, membranes, etc. must support and sustain the mechanical and thermo-mechanical loads associated with their use and operation. The micro-design of structural composites is a science, albeit an incomplete one.[2] The basic lesson we can learn from that field is that the morphology of the constituents (their shape, size, and relative shape and size) are critical in the determination of their properties and function. The size, shape, and arrangement of fibers, for example, in a fiber reinforced polymer matrix (from which modern airplanes are made) greatly influences the composite strength, stiffness, and life.[2] Indeed, these morphological details influence the local details of how the applied fields (mechanical stress and temperature, for example) are distributed and redistributed in the interior of the materials, resulting in very different response for different morphologies. The interaction of the heterogeneous constituent morphology with the applied fields also may introduce anisotropy in the response, by design in some cases, to achieve unique behavior. *So our second basic concept is that, in addition to interactions across material boundaries, constituent morphology, in every sense, is a fundamental concept in the creation of HeteroFoaMs.*

The third basic concept has to do with the unique role of transport in HeteroFoaMs. Batteries and separation membranes function by transporting mass by diffusion through dense solids. Fuel cells (and flow batteries, electrolyzers, etc.) also function by the transport of gasses or liquids through porosity in their solid phases. This mass transport is fully coupled to the balance of momentum, heat and charge that interactively controls the functionality of the energy device. It is the third salient, distinctive feature of HeteroFoaM materials that is responsible for the "FoaM" part of that acronym.

HETEROFOAM: ANALYSIS AND DESIGN

In the short space available here, and in the context of our focus on materials science and technology, we select only one material characteristic to illustrate the challenge of designing HeteroFoaM materials. We select the dielectric character of HeteroFoaM materials as our focus, because of its growing importance and especially because of the current lack of understanding and models available for the design of such materials for multiphysics functionality in the presence of dielectric effects. Our objective will be to identify the materials aspects of this problem (separate and apart from the usual discussions of electrochemical effects), and to present some new understandings of those aspects.

The dielectric properties of homogeneous and heterogeneous materials vary more greatly than many other physical constants, over tens of orders of magnitude. Moreover, they vary as a function of temperature, composition, order / disorder, and heterogeneity, and as a function of the frequency of excitation by an electric field.[3]

The dielectric response of materials is a fundamental property defined by the physical polarizations that develop in a material when an electric field is applied to the medium. Figure 2 illustrates some of the mechanisms by which that polarization is created, as a function of frequency of the excitation field. One can establish, for example, that at very high frequencies of the electric field, a coupling of the electric field oscillation to the relative local displacement of the positive nucleus and the negative electron cloud of the atoms in a gas create a capacitive polarization of the medium. For frequencies of 10^6, the polarization may result from interactions of the field with polar molecules such as water vapor. There is a rich literature that describes these broadband dielectric spectroscopy results.[4] At lower frequencies of excitation, surface and space charge mechanisms typically dominate.

Material dissipation of the energy of an input field is, in fact, a more general subject, and widely (and historically) investigated. Figure 3 illustrates a general listing of mechanisms associated with dissipation of input vibrational energy by various local (internal) mechanisms.

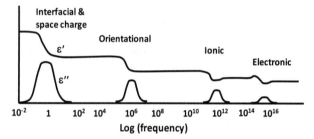

Figure 2. Real (top) and imaginary (bottom) dielectric permittivity as a function of frequency of excitation by electric field for a 'typical' dielectric material, with mechanisms associated with the response identified.

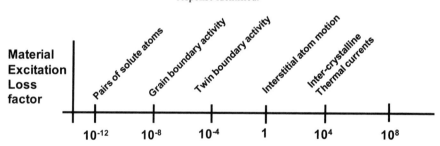

Figure 3. Material dissipation mechanisms in crystalline materials as a function of frequency of excitation by mechanical field.

In the present context, both of these phenomena result from an internal motion (associated with specific local mechanisms of interaction) induced by an external field. And both show that the interactions (especially at low frequencies of excitation) are affected by heterogeneity, e.g., interfaces, surfaces, and changes of properties across boundaries (as in polycrystalline materials).

Tuncer has studied the phenomenon of dielectric relaxation in dielectric mixtures.[5] The bulk of the literature he discusses focuses on effective property approximations that to not provide local information, and are not dependent on the local physics of the behavior in any direct way. Disordered low frequency dispersions in disordered binary dielectric mixtures (with equal volumes of each constituent) have also been studied by Tuncer.[6] Random, regular arrays of repeated local shapes were studied in that work. Indeed, essentially all of the effective property studies, classical as well as attempts at local solutions of the field equations, assume that the heterogeneity is uniform in local features.[7,8,9] This type of heterogeneity is generally not found in nature. In addition, the extensive work in the literature often assumes that there is a dielectric phase distributed in a matrix that is a better conductor. Some of the most interesting (and technologically important) situations are the reverse of that, with a 'better' conductor distributed in a more insulating medium.

In 2009, Fazzino, et al. published data that showed that polymer based composites, when they develop microcracking due to mechanical loading, change their dielectric response dramatically and definitively.[10,11] We have been able to determine two important aspects of that response. First, the dielectric response was associated with the moisture (and salt solutions) that collected in the microcracks. And second, the response is remarkably sensitive to the local morphology (including the topological connectivity) of the dielectric dispersed phase, e.g., the microcracks in this case.

More recently, Rabbi has reported results that show a clear association between the specific local morphology of a single-phase dielectric and its frequency dependent broadband response.[12] In that work there was a single material phase domain with a systematic variation of (only) internal morphology for which the Maxwell-Boltzman equation (below) was set on the local geometry and solved by COMSOL, to computationally calculate the frequency dependence of the dielectric response. To avoid impedance variations originating from material volume variation, each of the internal morphology cases considered had the same cross sectional area and thickness. Here COMSOL solves the following reduced Maxwell's equation:

$$-\nabla . d\big((\sigma + j\omega\varepsilon_0)\nabla V\big) = 0$$

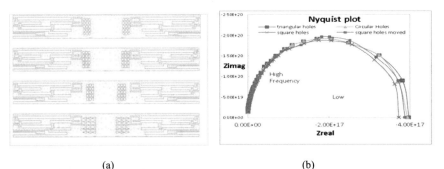

(a)   (b)

Figure 4. (a) Geometric variation in the form of rectangular, circular and triangular holes. (b) Impedance variation seen in Nyquist plot

where d is the material thickness, σ is the conductivity, ε is the permittivity, ω is the frequency of the incident field and j is the square root of (-1).

From Fig. 4 we can see that at lower frequency there is a distinct difference between impedance for different geometric variations, alone. Hence, even when all material properties and the volume is constant, the dielectric response, when the field equations are set on local morphology, can discriminate the details of the local morphology.

From our earlier discussion, it is clear that these results and observations are of importance to HeteroFoaM materials. HeteroFoaMs, or heterogeneous functional materials in which there is an active void phase, have exactly the combination of material features discussed above. They have multiple phases of different materials with complex (active) interfaces, and they have void phases which may be functional. Moreover, they are subject to microcracking when subjected to changes in applied thermal fields, for example, which recalls the work of Fazzino, et al.[10,11] In our brief discussion here, we will illustrate the importance of dielectric behavior in HeteroFoaM materials by considering a specific application of those materials, to solid oxide fuel cells.

The SOFCs we will discuss in this paper were 20 mm in diameter, button cells supplied by the ENrG Corporation, in a variety of formats and microstructures. The electrolyte was Ytria-stablized Zirconia (YSZ), and the cathode and anode were varied for our tests. As a baseline, single YSZ layers of the YSZ materials were characterized to establish the dielectric character of the materials involved. The dielectric properties obtained from those characterizations were then used in COMSOL, set on local geometries of the morphologies of interest, to obtain estimates of the dielectric response. The frequency response of those calculations was validated with a single phase, solid disc of YSZ material to confirm that predicted and observed broadband dielectric response was the same for dense, single phase solids. (More will be said of this work in another publication.)

Fig. 5 shows a single disc of YSZ, and a special thermal shock device designed and built in our laboratory that was used to subject that disc to a changes in temperature of about 400°C (from 400°C to 800°C) in seven minutes.

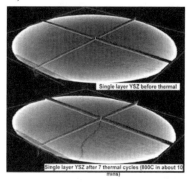
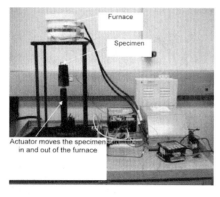

(a)   (b)

Figure 5. YSZ specimens (a) and thermal shock device (b) used to induce microcracking in SOFC button cells.

Surprisingly, the thermal shock of even this single dense layer of YSZ was able to induce not only microcracking but also macro-cracking as can be seen in the lower photograph of Fig. 5(a).

Figure 6 shows the broadband response of a fully operational SOFC button cell, again at room temperature, after one, two and three thermal shocks, using the equipment shown in Fig. 5. The changes in this response are definitive and distinctive, and consistent with explanations that we have previously published.[9,10,13]

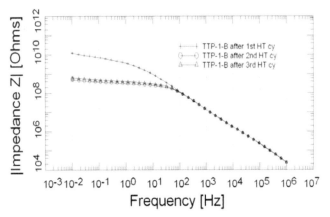

Figure 6. Broadband dielectric response of an SOFC fuel cell (at room temperature) before and after several thermal shocks.

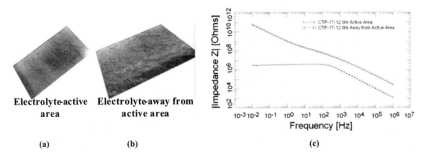

Electrolyte-active area   **Electrolyte-away from active area**

(a)   (b)   (c)

Figure 7. Micro-X-Ray tomography showing a crack in the active area (a), no cracking in the in-active area (b), and the broadband dielectric spectroscopy response of active and away from active regions of a SOFC button cell after 8 hr of operation (measured at room temperature) (c).

For the data shown in Fig. 7, Broadband dielectric spectroscopy (BbDS) measurements were carried out on areas inside and away from the active region of an SOFC button cell after operating the cell at 800°C for 8 hrs. These impedance data are measured in the absence of any electrochemical processes and at room temperature. It is postulated that during operation, material deterioration contributed to the formation of new morphological variations inside the SOFC structure that caused the observed variation in impedance response. From 3D X-ray tomography microscopy, the micro-structural differences were visualized and a representative virtual section of the electrolyte is shown in Fig 7.

There is a visible crack (damage) on the electrolyte near the electrolyte-electrode interface in the active area.

However, if micro/macro-cracking of that severity occurs in an operating SOFC, the cell is most likely no longer functional as a fuel cell. To examine the behavior that results from changes during the operation of an SOFC, a series of tests were conducted in which the broadband dielectric response was recorded at intermittent times during the test of an SOFC button cell. In all cases, the tests were done at open circuit voltage (OCV), with nitrogen flowing over the anode. For these tests a GAMRY™ EIS300 electrical impedance measuring device was used to record Impedance and Capacitance at 800°C. The GAMRY uses an AC input over a variable frequency range of 0.1Hz-1Mhz to measure the Impedance and capacitance response.

While the complete interpretation of the results of this experiment is still a work in progress, some of the features of those data are important for our present discussion.

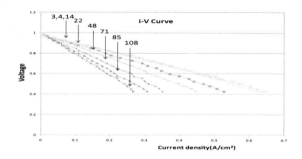

Figure 8. Voltage current (V-i) curve as a function of running time for SOFC test.

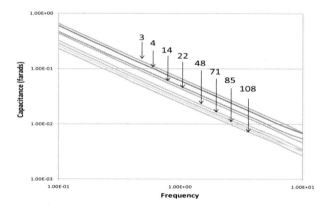

Figure 9. Broadband capacitance change with time (in hours) of operation of SOFC, measured at OCV.

Figure 8 shows the voltage-current (V-i) curves recorded (at OCV) after operating the cell for 3, 4, …108 hours. The decreasing slope of the V-i curve indicates that the Ohmic polarization of the cell is increasing with use, indicating that from an electrochemical standpoint the fuel cell is degrading. That degradation is caused by material state changes in the fuel cell materials and the degradation of the interfaces between materials in the fuel cell. (Those details will be discussed in another paper.) However, the degradation is systematic and significant. Figure 9 shows the capacitance recorded at OCV during each of the interruptions of the tests. These data clearly show that the total volumetric material capacitance of the SOFC decreased in proportion to the degradation of the power capability of the fuel cell during operation. We believe that this is the first reported correlation between the non-Faradaic material (only) dielectric response of fuel cell materials and their electrochemical function as a fuel cell. Moreover, the changes of the dielectric behavior shown in Fig. 9 are large, about half of an order of magnitude, perhaps a factor of five. The sensitivity of the method is remarkable and the systematic indication of degradation in proportion to total capacitance change is strongly indicated.

The data that are presented in Figs. 8 and 9 represent a fully functioning cell, albeit one which is degrading due to causes that are known to be electrochemical in nature. What happens when mechanical microcracking actually occurs in SOFCs? Individual microcracks in an operating cell are very difficult to detect experimentally. We have shown, above and in earlier publications, that dielectric changes are extremely sensitive to such cracks. But there is another interesting feature of such cracking, having to do with the alteration of the extant field in an SOFC, provided by the electrochemical potential. The OCV of a cell is typically calculated by Wagner's equation which corrects the Nernst voltage for polarization losses in the electrodes.

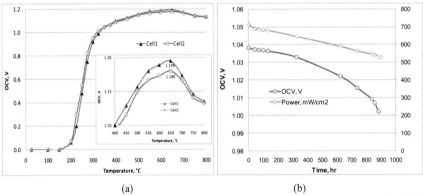

(a)   (b)

Figure 10. Typical SOFC OCV at various temperatures (a) and OCV changes over 900 hrs durability test (at 800°C and constant current). Both tests were using hydrogen.

Figure 10 (a) shows a typical example of SOFC OCVs at various temperatures. It can be seen that the maximum OCVs normally occur at temperatures between 550°C and 700°C for these cells. During long-term operation of an SOFC, the OCV also often decreases. For Example, Figure 10 (b) shows the results of a durability test of an SOFC that was operated for 900 hours in our laboratory. Over the 900 hours, the cell power degraded 25% while the OCV decayed by 3.5% from 1.038V to 1.002V. Figure 10 (b) also indicates that changes in the OCV of an SOFC may also be a sensitive measure of the fuel cell electrical performance.

In point of fact, we have no first principles method of calculating or predicting the observed OCV of an operating fuel cell that matches the laboratory value, especially after operation of the cell for some time.[14,15] One of the difficulties associated with this part of the physics is the prior lack of a closed form method for directly calculating the forces and torques on internal dielectric particles or features associated with a given applied electric field, i.e., the polarization. Recently, such a method has been presented for exactly that problem by one of the members of our group.[16,17] One of the interesting predictions of that work is that the OCV of an SOFC should be reduced by matrix cracking, if the reactant gases contain constituents that are ionic or have polar molecules (as is the case for SOFC operation). In simple form, the mechanical cracking creates regions of increased polarization (and capacitance) in the constant electric field created by the electrochemical potential of the reactants, reducing the electric field voltage across the cell. Therefore, it would appear that the broadband dielectric material response also provides a nondestructive method of detecting, quantitatively, that specific mechanism of degradation. It should be mentioned that reductions in OCV move the entire V-i curve downward, so that small changes in OCV represent significant changes in power output, as Fig. 10(b) confirms.

OPPORTUNITIES AND CHALLENGES

The present paper has discussed a new class of heterogeneous functional materials called HeteroFoaMs, which are the subject of the research being conducted in the DoE "HeteroFoaM Center," an Energy Frontiers Research Center involving seven universities and two national labs. We have defined this class of materials, highlighted several distinguishing features of that class, and presented one example of the uniqueness of the multiphysics that controls the collective functionality of these heterogeneous material systems. Examples of these features include the fact that multiphysics interactions of constituents across interfaces determine global functional properties; porosity in these solid-based systems is often another active functional phase; and morphology, in every context, is a fundamental part of the foundation of the control science needed to understand and synthesize, by design, these materials. Finally, we have elucidated some of the unique details of the dielectric behavior of HeteroFoaMs. For the technical application of SOFCs, for example, we showed that broadband dielectric (material) impedance is remarkably sensitive to internal microstructure and morphology, and that the non-Faradaic capacitance of an SOFC varies directly in proportion to the power output of the fuel cells for our tests. We also observed that OCV decreases systematically as the performance of the SOFC decreases, and predicted, from local solutions of polarization using our newly developed closed form analysis method, that micro- or meso-cracking is expected to have exactly that effect on this important performance parameter. This latter work opens a new area of science in what has previously been a highly empirical area of heuristic arguments based only on global thermodynamic reasoning.

There are many opportunities and challenges associated with these findings. Like the idea of heterostructures before it, the concept of HeteroFoaMs opens up a new design space for the creation of heterogeneous functional material systems for energy conversion and storage, solid membranes for technologies ranging from hydrogen separation in high temperature nuclear reactors to medical devices, and a wide range of power management and control devices. The key elements of the science foundation for these technical advances are: a) multiphysics analysis that correctly captures the coupled physics of the response to concomitant electric, thermal, mechanical and chemical fields; b) multiscale analysis that begins with atomic level property and functionality predictions, caries that functionality to meso-analysis wherein the field equations (balance equations for mass, momentum, energy and charge) are properly set on local morphology, with material coefficients and constitutive equations defined by atomic analysis to predict global functionality; and c) correct understandings and models of the interfaces, interphases, and surfaces between constituents, which control the interactions that so clearly define the collective functionality of these material systems. It is clear that there is much to be done in this field of effort, and that the results will impact our society in a variety of sectors, from energy technology, to transportation, to medical devices and health to name a few. The HeteroFoaM Center is dedicated to this pursuit.

ACKNOWLEDGMENTS

The authors gratefully acknowledge the financial support of the United States DARPA under contract ARL001-007 for the fuel cell testing effort conducted by Y. Du. The bulk of the work conducted and reported on HeteroFoaM materials by the remaining authors was supported by the Department of Energy under funding for an Energy Frontiers Research Center for heterogeneous functional materials (the HeteroFoaM Center), under Grant No. DE-SC0001061. The X-ray tomography work was conducted by P. Majumdar and R. Raihan with support from DARPA funding under grant number W91CRB-10-1-0007.

REFERENCES

[1] P. Robin, High Speed Heterostructure Devices, Cambridge University Press, 2001

[2] K..Reifsnider and S.W. Case, Durability and Damage Tolerance of Material Systems, John Wiley & Sons, New York, 2003.

[3] Peter Barber, Shiva Balasubramanian, Yogesh Anguchamy, Shushan Gong, Arief Wibowo, Hongsheng Gao, Harry J. Ploehn* and Hans-Conrad zur Loye, Polymer Composite and Nanocomposite Dielectric Materials for Pulse Power Energy Storage, *Materials* , 2, 1697-1733 (2009)

[4] Kremer, F. and Schonhals, A., Eds., Broadband Dielectric Spectroscopy, Springer, Berlin (2002)

[5] Tuncer, E. "Dielectric Relaxation in Dielectric Mixtures," Dissertation, Chalmers Tekniska Hogskola, Goteborg, 2001

[6] Tuncer, E., "Signs of Low Frequency Dispersions in Disordered Binary Dielectric Mixtures," J. Physics D: Applied Physics, 37, 334-342 (2004)

[7] Maxwell, J.C., "A Treatise on Electricity and Magnetism," Vol. 1, Oxford; Clarendon (1891)

[8] Sillars, R., J. Inst. Electr. Eng. 80, 378-94 (1937)

[9] Wagner, K.W., Archiv fur Electrontechnik II, 371-87 (1914)

[10] Reifsnider, K., Fazzino, P., and Majumdar, P., "Material State changes as a basis for Prognosis in Aeronautical Structures," J. Royal Aero. Soc., 113#1150, 789-98 (2009)

[11] Fazzino, P. and Reifsnider, K.L., "Electrochemical Impedance Spectroscopy Detection of Damage in Out-of -Plane Fatigued Fiber Reinforced Composite Materials," Appl. Composite Materials, 15(3), 127-38 (2008)

[12] Rabbi, F. and Reifsnider, K., "Relationship of Micro-Structure Morphology to Impedance in Heterogeneous Functional Materials," Proc. FuelCell2010, 33 #170, 393-397 (2010)

[13] Majumdar, P., Raihan, M., Reifsnider, K. and Rabbi, F., "Effect of Porouis electrode Morphology on Broadband Dielectric Characteristics of SOFC and Methodologies for Analytical Predictions," Proc. ASME 2011, 9th Fuel Cell Science Engineering and Technology Conf., Paper # 54946 (2011)

[14] Miyashita, T., "Unchanged OCV Requires Concepts Considering Electrode Degradation Using Sm-Doped Ceria Electrolytes, in SOFCs," Electrochemical and Solid-State Letters, 14#7, B66-69 (2011)

[15] Yoo, H.I., and Martin, M., "On the Path-Dependence of the Open-Cell Voltage of a Galvanic Cell Involving a Ternary or Multinary Compound with Multiple Mobile Ionic Species under Multiple Chemical Potential Gradients," Phys. Chem. Chem. Phys., 12, 14699-705 (2010)

[16] Liu, Q. ,"Physalis Method for Heterogeneous Mixtures of Dielectrics and Conductors: Accurately Simulating One Million Particles Using a PC," J. Comp. Physics 230, 8256-74 (2011)

[17] Liu, Q., Directly Resolving Particles in an Electric Field: Local Charge, Force, Torque, and Applications. *Int. J. Numer. Meth. Engng*, Accepted (2011)

STUDY ON HETEROPOLYACIDS/Ti/Zr MIXED INORGANIC COMPOSITES FOR FUEL CELL ELECTROLYTES

Uma Thanganathan
Research Core for Interdisciplinary Sciences (RCIS), Okayama University, Tsushima-Naka, Okayama, 700-8530, Japan

ABSTRACT

Analysis of complex impedance and ^{31}P NMR spectra were carried out on heteropolyacids and phosphosilicate based on inorganic materials TiO_2 and ZrO_2 mixed in the composites. ^{31}P NMR and impedance spectroscopies were used to study proton mobility in the composites. A high proton conductivities were found in the order of 10^{-2} was observed at room temperature. This inorganic ceramic composites were evaluated their properties by different characterizations. In this investigation, the properties of nuclear magnetic resonance and proton conductivity on ceramic composites were described and the results were presented.

INTRODUCTION

Recently much attention has been directed to proton conductors as materials for electrolyte applicable to fuel cells. A proton conducting membrane is an integral part of a fuel cell system. The protonic conductor has been studied for a large number of inorganic and polymer materials [1,2]. Since perfluoro sulfonic acid polymers such as Nafion®, well-known as the most popular solid-proton-conductors, have several disadvantages, e.g: high price and limited working temperature range below 100 °C. Nogami *et al.* proved that protons are the main charge carriers in phosphate glasses; it could be applicable for fuel cells [3,4]. A good proton-conducting membrane should combine high chemical, mechanical and thermal stabilities with reasonable electrochemical properties [5]. Protons in oxide glasses usually remain as X–OH groups (X = Si, P, etc) in a glass structure when glasses are prepared via a melting process in ambient atmosphere [6]. A significant contribution of residual P–OH groups in phosphate glasses to electrical conduction, that protons are the main charge carriers in phosphate glasses [7]. Furthermore, glasses with high concentrations of OH groups are potential proton conductors [8].

Proton transport includes transport of protons (H$^+$) and any assembly that carries protons (OH$^-$, H$_2$O, H$_3$O$^+$, NH$_4^+$, HS$^-$ etc). The transport of protons (H$^+$) between relatively stationary host anions is termed the Grothus or free-proton mechanism. Transport by any of other species is termed as a 'vehicle mechanism' [9]. In solids, vehicle mechanisms are usually restricted to materials with open structures (channels, layers) to allow passage of the large ions and molecules, while the Grothus mechanism requires close proximity of water molecules which are held firmly but are able to rotate. The classification of proton conductors according to the preparation method, chemical composition, structural dimensionality, mechanism of conduction etc. have been summarized in a comprehensive book on proton conductors [10]. From the literature of the existence of class of inorganic compounds which have very high protonic conductivity with respect to all other inorganic solid compounds: the heteropolyacids. In particular two of them, the phosphotungstic acid (PWA) and the phosphomolybdic acid (PMA), both containing in the crystalline 29 water molecules, show specific conductivities of 0.17 and 0.18 Ω^{-1}/cm, respectively, at room temperature [5].

Recently, incorporation of inorganic acids in polymer has been attempted for ionic conductive materials [11]. Heteropolyacid (HPAs) based proton conducting electrolytes have aroused a considerable interest for their protonic activity and oxygen affinity properties. PWA is known as a solid oxide cluster which has both high protonic conductivity and high solubility for water and alcohol, and many groups are used as doping inorganic solid acids. The acidic solid cluster of PWA has three protons in the unit structure, which dissociates in the water phase as oxionium ions as H_3O^+ or $H_5O_2^+$. The polyanion PWA clusters have been incorporated with inorganic silica frame work and never dissociate from the hybrid matrix. The basic structural unit of PWA is the Keggin anion $(PW_{12}O_{40})^{3-}$ which consists of the central PO_4 tetrahedron surrounded by four W_3O_{13} sets linked together through oxygen atoms. Each W_3O_{13} set is formed by three edge-sharing WO_6 octahedra. They form channels which can contain up to 29 water molecules in different hydrate phases. This variety leads to different protonic species and hydrogen bonds of different strength [12,13].

NMR has played a major role in the study of fast ionic/protonic conductors. Fast protons are conductors. There are a number of review articles [14,15], summarizing and critically examining the NMR studies of fast ionic conductors (FPCs), and many of them contain some discussion of fast protonic conductors. However, there are also aspects peculiar to FPCs, an important one being the influence of local dynamics. A more recent paper demonstrated proton exchange between water and sulfonic acid in a Nafion-type membrane [16]. Lim et al. [17] were reported that the physicochemical interaction of PMA doped with polymer by ^{31}P MAS NMR analysis. Our previous reports, we have described the properties of heteropolyacid and phosphosilicate mixed with inorganic TiO_2/ZrO_2 ceramic composite such as structural and proton conductivity [18–20]. In this work, new class, proton conducting composite membranes (CM) are being designed and developed. The structural properties of PWA/PMA and TiO_2/ZrO_2 mixed phosphosilicate ceramic composites are characterized by ^{31}P MAS NMR. Impedance spectroscopies were used to study proton mobility in hydrated samples.

EXPERIMENTAL SECTION

All chemicals were used as obtained. Metal alkoxides and heteropolyacid were used to synthesize the $PWA/PMA/TiO_2/ZrO_2/P_2O_5/SiO_2$ inorganic composite membranes. They were prepared by following the sol-gel method reported earlier [18–20]. The porous ceramic composites were heat-treated at 600 °C.

Magic-angle spinning nuclear magnetic resonance (^{31}P NMR) spectra were obtained using an NMR spectrometer (model Unity 400, Varian) operating at 161.906 MHz, spinning speed 5 kHz; delay time 10s; pulse length 8.8 µs. The chemical shifts were measured relative to an 85% H_3PO_4 solution. The powder samples were stored in a closed vessel and dried for at least 1 h prior to before this measurement. The electrical conductivities (σ) on phosphosilicate based ceramic composites $(PWA/PMA/TiO_2/ZrO_2/P_2O_5/SiO_2)$ consisting heteropolyacids (PWA and PMA) and oxides (TiO_2 and ZrO_2) were determined at room temperature from Cole–Cole plots by an ac method using an impedance frequency analyzer (SI–1260, Solartron). The electrical conductivity was obtained from the intersection of the Cole–Cole semicircle plot with the real axis. A frequency range of 0.1 Hz to 1 MHz with a peak-to-peak voltage of 10 mV was used for the impedance measurements. Impedance spectra were recorded in the frequency range from 1 MHz to 0.1 Hz (10 points per decade) with a voltage amplitude 10 mV in humid atmosphere. Spectra were simulated by means of Z view software (Scribner Associates, Version 2.1). The proton conductivity σ of the membranes was determined from the bulk resistance R, the thickness d, and the sample area A.

RESULTS AND DISCUSSION

^{31}P NMR study and a mechanism of proton conduction on $PWA/PMA/TiO_2/ZrO_2/P_2O_5/SiO_2$ ceramic composites were proposed. The ^{31}P MAS NMR spectra of $PWA/P_2O_5/SiO_2$, $PMA/P_2O_5/SiO_2$, $PWA/PMA/P_2O_5/SiO_2$, $PWA/TiO_2/P_2O_5/SiO_2$, $PWA/ZrO_2/P_2O_5/SiO_2$, $PMA/TiO_2/P_2O_5/SiO_2$, $PMA/ZrO_2/P_2O_5/SiO_2$, $PWA/PMA/TiO_2/P_2O_5/SiO_2$ and $PWA/PMA/ZrO_2/P_2O_5/SiO_2$ inorganic composite samples were showed a large resonance in the positive side, their datas are listed in Table 1.

Table I. ^{31}P MAS NMR spectra of PWA/PMA/TiO$_2$/ZrO$_2$/P$_2$O$_5$/SiO$_2$ ceramic composites

No	Samples	Inorganic composites	Chemical shift/ppm
1	A	PWA/P$_2$O$_5$/SiO$_2$	7.62
	B	PWA/TiO$_2$/P$_2$O$_5$/SiO$_2$	4.82
	C	PWA/ZrO$_2$/P$_2$O$_5$/SiO$_2$	4.6-4.7
2	A	PMA/P$_2$O$_5$/SiO$_2$	6.49
	B	PMA/TiO$_2$/P$_2$O$_5$/SiO$_2$	5.37
	C	PMA/ZrO$_2$/P$_2$O$_5$/SiO$_2$	5.15
3	A	PWA/PMA/P$_2$O$_5$/SiO$_2$	6.98
	B	PWA/PMA/TiO$_2$/P$_2$O$_5$/SiO$_2$	5.46
	C	PWA/PMA/ZrO$_2$/P$_2$O$_5$/SiO$_2$	4.6-4.7

Hydrogen bonding plays an important role in terms of structure and proton transport in ceramic composites. The physicochemical interaction of heteropolyacids with phosphosilate and TiO$_2$/ZrO$_2$ oxides was further confirmed by the ^{31}P NMR analysis. As shown in figures 1 (A, B and C), PWA/P$_2$O$_5$/SiO$_2$ and PWA/TiO$_2$/P$_2$O$_5$/SiO$_2$ and PWA/ZrO$_2$/P$_2$O$_5$/SiO$_2$ samples showed the singlet, the chemical shift observed for PWA–P$_2$O$_5$/SiO$_2$ was 7.62 ppm, PWA/TiO$_2$/P$_2$O$_5$/SiO$_2$ was 4.82 ppm and PWA/ZrO$_2$/P$_2$O$_5$/SiO$_2$ was 4.60-4.69. From these result, we have observed the ^{31}P chemical shift for PWA/P$_2$O$_5$/SiO$_2$ is slightly different from PWA/TiO$_2$/ZrO$_2$/P$_2$O$_5$/SiO$_2$ [figures 1 (B and C)] heteropoly-phosphosilicate inorganic composites, although both samples exhibited the singlet. Similar trends were observed in figures 2 and 3, respectively.

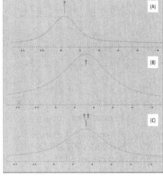

Chemical shift (ppm)

Figure 1. ^{31}P MAS NMR spectra of ceramic composites: (A) PWA/P$_2$O$_5$/SiO$_2$, (B) PWA/TiO$_2$/P$_2$O$_5$/SiO$_2$ and (C) PWA/ZrO$_2$/P$_2$O$_5$/SiO$_2$. All spectra acquired at room temperature control and with spinning frequencies of 5 kHz.

It is known that the ^{31}P chemical shifts for HPA in the solid-state NMR strongly depend on the identity and the number of coordinated organic molecule [21]. The position shifts are more positive and the line width becomes progressively narrower and the intensity is higher for PWA containing the phosphosilciate ceramic samples (figure 2) than PMA contains ceramic samples (figure 1). In our previous paper, we have reported that ^{31}P NMR spectrum for the P$_2$O$_5$/ZrO$_2$/SiO$_2$ system [22], and observed only one sharp signal at 0 ppm, indicating that not all the P^{5+} ions are polymerized but some remain as monomers to form PO(OH)$_4$ and other hand P^{5+} ions bound to one bridging oxygen.

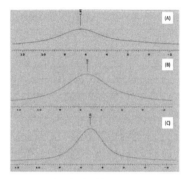

Chemical shift (ppm)

Figure 2. ^{31}P MAS NMR spectra of ceramic composites: (A) PMA/P$_2$O$_5$/SiO$_2$, (B) PMA/TiO$_2$/P$_2$O$_5$/SiO$_2$ and (C) PMA/ZrO$_2$/P$_2$O$_5$/SiO$_2$. All spectra acquired at room temperature control and with spinning frequencies of 5 kHz.

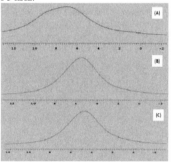

Chemical shift (ppm)

Figure 3. ^{31}P MAS NMR spectra of ceramic composites: (A) PWA/PMA/P$_2$O$_5$/SiO$_2$, (B) PWA/PMA/TiO$_2$/P$_2$O$_5$/SiO$_2$ and (C) PWA/PMA/ZrO$_2$/P$_2$O$_5$/SiO$_2$. All spectra acquired at room temperature control and with spinning frequencies of 5 kHz.

Also we were studied the ^{31}P NMR spectrum for the P$_2$O$_5$/ZrO$_2$/SiO$_2$ system [18], and observed the NMR spectral features indicate that the Ti^{5+} ions act with the P^{5+} ions. By comparison, the phosphosilicate composites without content heteropolyacids, the chemical signals were appeared both positive and negative sides, but in the heteropoly-phosphosilicate composites showed the ^{31}P NMR spectrum signals are shifted to positive sides only. It was believed that the heteropolyacid ions are attributed to the ceramic composite network. Figure 3 (A), PWA/PMA/P$_2$O$_5$/SiO$_2$ ceramic composite shows a broad shift at 6.98 ppm, PWA/PMA/TiO$_2$/ZrO$_2$/P$_2$O$_5$/SiO$_2$ (sample A and B) shows the signal is lower than sample C, although singlet in the positive side. Keggin structure was partially overplayed by the Si–O frame vibrations [23]. Ti–OH$_2^+$ groups are able to strongly attract the HPAs through electrostatic interactions [19].

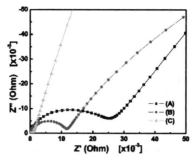

Figure 4. Cole–cole plots of complex impedances measured on ceramic composite memrbanes at room temperature: (A) $PWA/P_2O_5/SiO_2$, (B) $PWA/TiO_2/P_2O_5/SiO_2$ and (C) $PWA/ZrO_2/P_2O_5/SiO_2$.

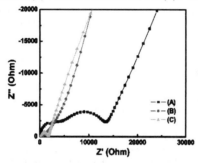

Figure 5. Cole–cole plots of complex impedances measured on ceramic composite memrbanes at room temperature: (A) $PMA/P_2O_5/SiO_2$, (B) $PMA/TiO_2/P_2O_5/SiO_2$ and (C) $PMA/ZrO_2/P_2O_5/SiO_2$.

Figures 4, 5 and 6 shows Cole–Cole plots of the complex impedance measured on $PWA/PMA/TiO_2/ZrO_2/P_2O_5/SiO_2$ inorganic composites at room temperature, the obtained plots showed parts of one or two arcs. It's not showed any much difference at room temperature. The value of σ is in the order of 10^{-2} S/cm at room temperature, it was estimated from the x-intercept of the extrapolation of the arcs. These ceramic composites have been small pores; it acts to accelerate the proton hopping, resulting high proton conductivity at room temperature. The motion of the water molecules is restricted by the small sized pores. We have described how the pore properties affect the proton conduction [18,19]. Proton conductivity measurements to assess the contribution of each proton transport mechanism in composite heteropolyacid-ceramic composite electrolytes. The high proton conductivity is attributed to the glass network contains HPA (PWA/PMA) acid ions as a proton donor and strongly hydrogen-bonded hydroxyl groups combined to P–O–Si bonds as proton conduction paths. Phosphosilciate composites have been pore size in the range from 1–10 mm. The pore size distribution gives rise to a distribution of spin-lattice relaxation rates than can potentially be extracted from the non-exponential, longitudinal magnetization decay [24–26].

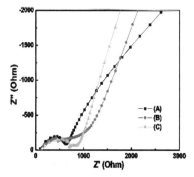

Figure 6. Cole–cole plots of complex impedances measured on ceramic composite memrbanes at room temperature: (A) PWA/PMA/P$_2$O$_5$/SiO$_2$, (B) PWA/PMA/TiO$_2$/P$_2$O$_5$/SiO$_2$ and (C) PWA/PMA/ZrO$_2$/P$_2$O$_5$/SiO$_2$.

This present conductivity results are not higher than previous reports [27]. A proton conductivity of ~1 S/cm was achieved at 30 °C for PWA/PMA/P$_2$O$_5$/SiO$_2$ inorganic composites. This is the highest conductivity obtained for a glass. These differences are according to the preparation method, composition, structural dimensionality and conductivity mechanism. It is well known that the protonic conductivity of heteropolyacids is strictly related to the number of water molecules coordinated to the Keggin anions. Furthermore, ceramic composites with high concentrations of OH groups are potential proton conductors. The principal proton conduction mechanism involves proton transfer between adjacent OH$^-$ and O^{2-} and OH$^-$ reorientation (Grotthuss mechanism) rather than OH$^-$ diffusion as sometimes emphasized [28,29]. The observation from a comparison of NMR and conductivity results, heteropoly-phosphosilicate ceramic composite samples were showed similar trends for both measurements, reveals that proton mobility would be associated with that of water molecules at low temperature.

CONCLUSION
^{31}P NMR spectrum of the heteropoly-phosphosilicate ceramic composites containing TiO$_2$ and ZrO$_2$ in silica network indicated features of Ti–O–Si and Zr–O–Si linkages; also PWA and PMA were partially overplayed by the Si–O frame vibrations. ^{31}P NMR measurements exhibited remarkable changes in chemical shift for the signals for all the ceramic composites at room temperature. Sol-gel derived porous ceramic composites were yielded high proton conductivities at room temperature. These inorganic composite membranes are suitable candidate as an electrolyte for fuel cell applications.

ACKNOWLEDGEMENTS
This work was financially supported by the Ministry of Education, Sport, Culture, Science and Technology (MEXT) and the Special Coordination Funds for Promoting Sciences and Technology of Japan. Thank to Prof. M. Nogami, Department of Material Science and Engineering, Nagoya Institute of Technology, Japan, for his discussion in the experimental section.

*Corresponding author. Tel.: +81 86 251 8706; fax: +81 86 251 8705
E-mail: umthan09@cc.okayama-u.ac.jp

REFERENCES

[1]K. D. Kreuer, "Proton conductivity. Materials and applications", *Chem Mater.* **8**, 610–641 (1996).

[2]N. Giordano, P. Staiti, A. S. Arico, E. Passalacqua, L. Abete and S. Hocevar, "Analysis of the chemical cross-over in a phosphotungstic acid electrolyte based fuel cell", *Electrochim Acta.* **42**, 1645–1652 (1997).

[3]Y. Abe, G. Li, M. Nogami and T. Kasuga, "Superprotonic conductors of glassy zirconium phosphates", *J. Electrochem Soc.* **143**, 144–147 (1996).

[4]M. Nogami, R. Nagao, K. Marita and Y. Abe, "Fast proton-conducting $P_2O_5/ZrO_2/SiO_2$ glasses", *Appl. Phys Lett.* **71**, 1323–1325 (1997).

[5]R.A. Lemons, "Fuel cells for transportation", *J. Power Sources*, **29**, 251–264 (1990).

[6]K. Kawamura, H. Hosono, H. Kawazoe, N. Matsunami and Y. Abe, "Large Enhancement of Electrical Conductivity in Magnesium Phosphate Glasses by Ion Implantation of Proton", *J. Ceram. Soc. Jpn.* **104**, 688–690 (1996).

[7]H. Hosono, T. Kamae and Y. Abe, "Electrical conduction in magnesium phosphate glasses containing heavy water", *J. Amer. Ceram. Soc.* **72**, 294–297 (1989).

[8]K.D. Kreuer, "On the development of proton conducting materials for technological applications", *Solid State Ionics,* **97**, 1–15 (1997).

[9]K.D. Kreuer, A. Rabenau and W. Weppber, "Vehicle mechanism, A new model for that interpretation of the conductivity of fast proton conductors", *Angew.chem.Int. Ed. Engl.* **21**, 208–209 (1982).

[10]H. Ikawa, (Ed.) P. Colamban, Proton conductors, Cambridge University Press, p. 190 (1992).

[11]J.M. Amarilla, R.M. Rojas, J.M. Rojo, M.J. Cubillo, A. Linares and J.L. Acosta, "Antimonic acid and sulfonated polystyrene proton-conducting polymeric composites", *Solid State Ionics,* **127**, 133–139 (2000).

[12]U. Mioc, M. Ddavidovic, N. Tjapkin, Ph. Colomban and A. Novak, "Equilibrium of the protonic species in hydrates of some heteropolyacids at elevated temperatures", *Solid State Ionics,* **46**, 103–109 (1991).

[13]C. Rocchiccioli-Deltcheff, M. Fournier, R. Franck and R. Thouvenot, "Vibrational investigations of polyoxometalates. 2. Evidence for Anion-Anion Interactions in Molybdenum (VI) and Tungsten (VI) compounds related to the Keggin Structure", *Inorg. Chem.,* **22**, 207–216 (1983).

[14]J.B. Boyce and B.A. Huberman, "Superionic conductors-Transitions, structures, dynamics", *Physics Reports-Review section of Phys Letts.,* **51**, 189–265 (1979).

[15]J.L. Bjorkstam, "1st and 2nd order effects in NMR spectral averaging", *J. Mol. Struct.* **111**, 135–150 (1983).

[16]G. Ye, N. Janzen and G.R. Goward, "Solid-state NMR study of two classic proton conducting polymers: Nafion and sulfonated poly(etheretherketone)s", *Macromolecules*, **39**, 3283–3290 (2006).

[17]S.S. Lim, G.L. Park, I.K. Song and W.Y. Lee, "Heropolyacid (HPA)-polymer composite films as catalytic materials for heterogeneous reactions", *J. Molecular Catalysis A: chemical*, **182/183**, 175–183 (2002).

[18]T. Uma and M. Nogami, "Influence of TiO_2 on proton conductivity in fuel cell electrolytes based sol-gel derived P_2O_5-SiO_2 glasses, *J. Ion-Cryst Solids* , **351**, 3325–3333 (2005).

[19]T. Uma and M. Nogami, "Development of new glass composite membranes and their properties for low temperature H_2/O_2 fuel cells", *ChemPhysChem.* **8**, 2227–2234 (2007).

[20]T. Uma and M. Nogami, "Structural and textural properties of mixed PMA/PWA glass membranes for H_2/O_2 fuel cells", *Chem. Mater.,* **19**, 3604–3610 (2007).

[21]T. Okuhara, N. Mizuno and M. Misono, "Catalytic chemistry of heteropoly compounds", *Adv. Catal.* **41**, 113–252 (1996).

[22]M. Nogami, Y. Goto and T. Kasuga, "Proton conductivity in Zr^{4+}-ion-doped P_2O^5-SiO_2 porous glasses", *J. Am. Ceram.Soc.,* **86**, 1504–1507 (2003),

[23]B.B. Bardin and R.J. Davis, "A comparison of cesium-containing heteropolyacid and sulfated zirconia catalysts for isomerization of light alkanes", *Topics Catal.* **6**, 77–86 (1998).

[24]J.P. Korb, B. Sapoval, C. Chachaty, and A.M. Tistchenko, "Nuclear relaxation and fractal structure of a cross linked polymer", *J. Phys. Chem.* **94**, 953–958 (1990).

[25]K.S. Mendelson, "Nuclear magnetic-relaxation in fractal pores", *Phys. Rev.*, **B34**, 6503–6505 (1986).

[26]J.P. Korb and C. Chachaty (Proc. Of the GERM XI, Draveil, France (1990); C. Chachaty J.P. Korb, J.R.C. Van der Maarel, W. Brass and P. Quinn, *Phys. Rev.* **B44**, 4771–4793 (1991).

[27]T. Uma and M. Nogami, "A highly proton conducting novel glass electrolyte", *Anal.Chem.*, **80**, 506–508 (2008).

[28]T. Yajima, H. Iwahara and H. Uchida, "Protonic and oxide ionic-conduction in BACE03-based ceramics-effect of partial substitution for BA in BACE0.9D0.103-Alpha with CA", *Solid State Ionics*, **47**, 117–124 (1991).

[29]H. Iwahara, H. Uchida and K. Morimoto, "High temperature solid electrolyte fuel cells using perovskite type oxide based on BACE03", *J. Electrochem. Soc.*, **137**, 462–465 (1990).

Energy Storage: Materials, Systems and Applications

FATIGUE TESTING OF HYDROGEN-EXPOSED AUSTENITIC STAINLESS STEEL IN AN UNDERGRADUATE MATERIALS LABORATORY

Patrick Ferro, John Wallace, Adam Nekimken, Travis Dreyfoos, Tyler Spilker, Elliot Marshall
Gonzaga University
Spokane, WA, USA

ABSTRACT

Hydrogen storage designs are analyzed and tested in an undergraduate materials lab. Part of the focus is on possible degradation of mechanical properties of stainless steel exposed to hydrogen for long periods of time. To test for possible degradation of mechanical properties, samples of austenitic stainless steel are exposed to hydrogen and tested in fatigue. An experimental approach involving intermittent hydrogen exposure and resumed fatigue testing is under development as a possible alternative to in situ hydrogen testing. Future mechanical tests for intermittently charged specimens include tensile, impact and load relaxation testing.

INTRODUCTION

The literature pertaining to fatigue of steels and stainless steel alloys under hydrogen embrittlement conditions is an active area of research around the world. For example, a recent search of the literature using Google Scholar indicates more than 600 published articles and patents contain the keyword 'hydrogen embrittlement' since 2011[1]. Some of the driving forces behind the publication activity in this area include hydrogen embrittlement concerns in petroleum engineering, civil engineering, aerospace engineering and chemical engineering applications, for example. Additionally, the development of new materials for alternative energy-based vehicles and devices has also driven some of the recent activity in the investigation of hydrogen embrittlement. Hydrogen development and applications research is systematically investigated and published at a continuously updated DOE website[2]. A background in the mechanisms of hydrogen embrittlement may be found in classical reviews by authors including H.G. Nelson[3]. Active hydrogen embrittlement and fatigue research is being performed at Sandia National Laboratory[4,5], the University of Illinois at Urbana/Champaign[6,7], Kyushu University[8,9], Zhejiang University[10] and other institutions. Zheng et al. have provided a recent compilation of hydrogen embrittlement mechanical properties testing capabilities[10].

One of the challenges in performing mechanical properties tests on candidate materials for use in the hydrogen infrastructure is creating test apparatus that are hydrogen-compatible. For example, to investigate hydrogen embrittlement mechanisms for new materials, it may be necessary to have hydrogen at the tip of an advancing crack. This can be achieved by creating a test rig inside a hydrogen pressure chamber, or may be achieved by precharging hydrogen specimens at high pressure and temperature prior to testing in air. Each of these techniques has its limitations based on diffusivity of the material, safety, cost and other factors. The current work seeks to develop an intermittent hydrogen exposure test protocol to enable the fatigue testing of candidate materials for hydrogen exposure conditions. Other objectives of the hydrogen research program, at Gonzaga University, are to develop a hydrogen storage testing laboratory capability, and to introduce undergraduate engineering students to hydrogen testing and engineering.

BACKGROUND

The major types of materials that are considered for hydrogen containment and transportation are steels (including several grades of steel designed for petroleum service), austenitic stainless steels (including Types 304 and 316 nominally), brass alloys, aluminum alloys and fiber composites. Effectively studying hydrogen embrittlement phenomena in each of these respective material types requires exposing materials to hydrogen. Part of the experimental limitations include a consideration of the respective diffusivities of each of these material types, which likely influences the time that hydrogen is resident in a given material. Materials with high hydrogen diffusivities are expected to have low hydrogen concentration levels from outgassing if removed from a hydrogen environment.

One of the mechanisms by which hydrogen affects the mechanical properties of metal is by a dislocation mechanism. Hydrogen may collect near dislocations, which may then inhibit dislocation movement. Reduced dislocation movement may lead to localized reduction in ductility, which can lead to crack tip sharpening, manifesting in reduced bulk ductility and enhanced crack tip propagation.

Studying a material's relative propensity for hydrogen embrittlement via dislocation mechanism requires hydrogen to be continually present at an advancing crack tip. One method for ensuring this condition is to test the sample under a constantly applied source of hydrogen (e.g. in a hydrogen pressure vessel). Embrittlement failure by an externally applied source of hydrogen is an example of External Hydrogen Embrittlement (EHE).

Some alloys, when exposed to hydrogen for a sufficient period of time and under the right pressure and temperature conditions, absorb hydrogen interstitially until a saturation level is reached. If the hydrogen diffusivity in these alloys is sufficiently low, atomic hydrogen will remain in the metal even after it is removed from the hydrogen environment. Because hydrogen is infused throughout the material, including at dislocation sites and at advancing crack tips, the failure phenomenon for the material may be adversely affected. If the infused hydrogen inhibits dislocation movement and prevents crack tip blunting, the material will be embrittled by the internal source of hydrogen. This type of embrittlement phenomenon is an example of Internal Hydrogen Embrittlement (IHE).

Other manifestations of EHE and IHE may be possible, including failure phenomena associated with electrolytically generated hydrogen (which may occur during welding). Also, other failure mechanisms besides a dislocation mechanism may be possible but are not considered in the above discussion. Other possible types of failures including blistering and hydride formation are not considered in the present investigation.

Austenitic stainless steel (including Type 304, nominally) is an example of a material type that may be selected for hydrogen service. The combination of reasonable strength and ductility make it a suitable candidate material for engineering structures that are expected to perform in hydrogen gas conditions. Other material types that may be selected include aluminum alloys, brass and copper alloys and certain types of steel. A common property of these material types is good ductility, which mitigates potential embrittlement. In the present investigation, Type 304 stainless was selected for fatigue testing. The material was supplied as annealed sheets of 0.9 mm thickness.

Working with austenitic stainless steel specimens in the present investigation allows for studying internal hydrogen embrittlement, since the diffusivity of hydrogen in austenitic stainless is relatively low. With a relatively low hydrogen diffusivity, the specimens can be pre-charged with hydrogen and then tested outside the hydrogen atmosphere.

The diffusivity of hydrogen does not allow for appreciable outgassing during the time that the specimens are outside of the hydrogen atmosphere. Table 1 lists the diffusivities of hydrogen in different host materials, as an example of how low the diffusivity for hydrogen is in austenitic stainless steel[4]. The data in Table 1 is from San Marchi et al.[4], and was compiled from several sources[11-14]. The fifth column in the Table gives a relative distance, x, to show the effect of diffusion coefficient on hydrogen transport in each of the respective materials.

Table 1.   Hydrogen diffusivity parameters for some structural metals[4]

Material	D_o (m^2 s^{-1})	H_D (kJ mol^{-1})	D_{298K} (m^2 s^{-1})	$x=2\sqrt{Dt}$ (mm)	reference
austenitic stainless steel	5.76 x 10^{-7}	53.62	2.3 x 10^{-16}	0.17	[11]
low-alloy steel	3.5 x 10^{-7}	7.95	1.4 x 10^{-8}	1300	[12]
aluminum	1.75 x 10^{-8}	16.2	2.5 x 10^{-11}	56	[13]
copper	1.06 x 10^{-6}	38.5	1.9 x 10^{-13}	4.9	[14]

The fatigue samples used in the present investigation are relatively thin (0.9 mm). One of the implications of a relatively thin sample is that hydrogen may diffuse out of a precharged sample to cause appreciable hydrogen loss in the sample. The hydrogen concentration profile through the thickness of a sample can be estimated using a solution to Fick's second law, given in equation (1).

$$c(x) = 0.159 \, \text{erf}\left(\frac{x}{2\sqrt{Dt}}\right) \tag{1}$$

Equation 1 calculates the estimated hydrogen profile through the thickness (x) as a function of time (t). D is the diffusivity of hydrogen, given in Table 1. The coefficient term, 0.159 mol l^{-1}, is the estimated full charge capacity of hydrogen in austenitic stainless and is based on the material absorbing 2% by weight hydrogen. The model that was created to estimate the hydrogen concentration profile in fatigue samples assumes that samples start out fully charged (to 0.159 mol l^{-1}). Hydrogen begins to diffuse out when removed from the hydrogen charging atmosphere. The local concentration of hydrogen given by equation 1 is a function of distance from the surface (x), diffusivity and time. One of the assumptions of estimating the concentration profile using equation 1 is that diffusion is the rate-limiting step for hydrogen outgassing, and that surface reactions are occurring faster than the diffusion rate of hydrogen atoms through the metal to the surface.

Figure 1 shows the estimated hydrogen concentration profile across the fatigue specimen thickness for an austenitic stainless specimen. Profiles for the hydrogen concentration for 100h, 1000h and 10000h after removal from the hydrogen charging atmosphere are given. The profiles show that even after 10000h, most of the specimen is still infused with hydrogen, indicating that internal hydrogen embrittlement may still be investigated with samples that are precharged.

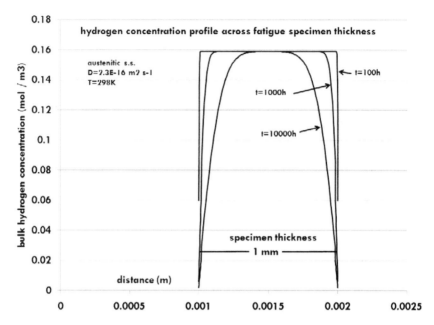

Figure 1.   Estimated hydrogen concentration profile across a fatigue specimen. The data shows the calculated hydrogen profile for an austenitic stainless sample at 100h, 1000h and 10000h after removal from a hydrogen atmosphere. The data indicates that appreciable hydrogen loss does not occur due to the relatively low diffusivity of hydrogen in austenitic stainless steel.

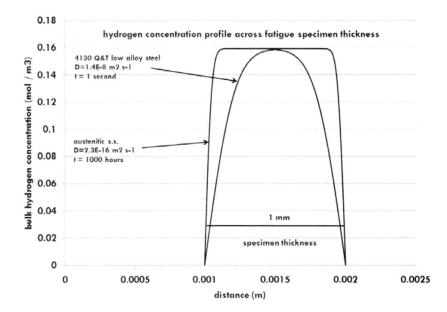

Figure 2.   Estimated hydrogen concentration profiles for steel and austenitic stainless steel, showing calculated hydrogen concentration after removal from a hydrogen environment as a function of position through the specimen thickness. Concentration profiles for 1 second (steel) and 1000 hours (austenitic stainless) after removal from a hydrogen atmosphere are shown. The hydrogen loss in the steel specimen is much greater than that for the austenitic stainless specimen, due to the relatively high hydrogen diffusivity in steel.

Hydrogen diffusion in steel is expected to be slower than that for austenitic stainless. Figure 2 compares estimated hydrogen concentration profiles for steel and austenitic stainless, using equation 1 and the diffusion coefficients form Table 1. The calculated data in figure 2 shows the estimated concentration profile through a fatigue specimen of 1mm thickness. Because the diffusivity of hydrogen in steel is eight orders of magnitude faster than that for austenitic stainless, the profiles shown in figure 2 are for different times. The concentration profile for steel is for only one second after removing the sample from the hydrogen atmosphere, compared with after 1000 hours, for the austenitic sample.

PROCEDURE
    The bending fatigue tests that were performed were using a VSS-40H bending fatigue testing machine from Fatigue Dynamics (Walled Lake MI) with a shutdown controller. The shutdown threshold percentage was set while the specimen was initially cycling at a preset cycling rate. The shutdown threshold percentages that were used for the experiments were 90% or 95% of the full initial applied bending stress. For relatively low initial applied stresses, relatively higher threshold percentages were used. The cycling frequency was between 300 and 350 cycles per minute. Failure

was defined as when the stress necessary to sustain bending dropped to the set threshold percentage of the initial applied bending load. Prior to installing a specimen into the machine, the displacement of the rotating arm on the fatigue tester was set for a given displacement (and applied stress) of the fatigue specimen. After the specimen failed, it was removed and the location of the failure was identified by measuring the distance from the start of the specimen radius to the middle of the failure location on specimen. Further description of the experimental equipment and procedures may be found in a paper by Ferro[15].

Fatigue specimens were plasma cut from 30.5 cm x 30.5 cm sheets of Type 304 stainless stee, and belt-ground along the periphery to remove most of the slag. The sheets were 0.9 mm (0.035 inches) thick, and in the annealed condition from Alcobra Metals (Spokane, WA). Figure 3 shows a drawing of the fatigue specimens.

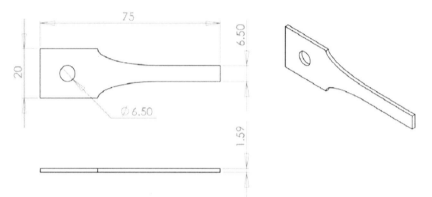

Figure 3.   Sketch of the bending fatigue specimen design that was used for the experiments. Dimensions shown are in mm. The thickness of the specimen shown above is 1.59 mm (0.064 in. The thickness for the specimens used in the present experiments was 0.9 mm (0.035 in).

Some of the samples were highly precharged with hydrogen at Sandia National Laboratory prior to testing. These samples were sent to Sandia in the plasma-cut and edge belt ground condition, where they were exposed to hydrogen pressures up to 138 MPa (20000 psi), at temperatures as high as 300°C (575°F) for a period of time of up to 15 days[5]. The typical hydrogen concentration in 300 series stainless steels after similar conditions is approximately 140 ppm, by weight[5]. Other hydrogen charging protocols are described by Mine et al.[9]

Some of the samples were exposed to 0.1 MPa hydrogen intermittently during testing at the materials laboratories at Gonzaga University. The samples were placed into a flask, and exposed to 0.1 MPa hydrogen for a minimum of 12 hours prior to fatigue testing. The samples were cycled in the fatigue tester for 6000 cycles and placed back into the flask for another 12 hours (minimum) hydrogen exposure. The source of the hydrogen was from portable hydrogen storage canisters. To create a hydrogen atmosphere in the flask, hydrogen from the canisters was discharged into the flask for a minimum of five minutes.

Bending stress was estimtaed by using elasticity equations for a cantilevered beam. The specimen was approximated as a beam of rectangular cross-section with an average width of 13.5 mm. The bending stress was calculated at a distance of 1 cm from the clamped end of the specimen, which was the average location of failure. Equations 2 and 3, from Mechanics of Materials texts[16], give the deflection as a function of distance x (equation 2) and the maximum deflection (equation 3) as a function of applied force, material stiffness, length and area moment of inertia of the beam cross section. Figure 4, from Beer et al[16], shows an example of a cantilevered beam and how force was estimated based on deflection.

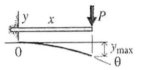

Figure 4.   Schematic diagram of an elastically deformed beam, showing force P required to elastically deflect the tip a distance y_{max}[16].

$$y = \frac{Px^2}{6EI}(3L - x) \tag{2}$$

$$y_{max} = -\frac{PL^3}{3EI} \text{ at x = L} \tag{3}$$

To calculate stress, equation 2 was used to find the force to achieve a given deflection. Once the force (P) was known, the maximum bending stress at the location on the beam where failure most often occurred on specimens was calculated using equation 3.

RESULTS

Figure 5 shows the number of cycles to failure (N_f) at a maximum applied bending stress (S_{max}) of 183 MPa. The data shown in figure 5 are for samples that were highly precharged with hydrogen at Sandia (circles), intermittently exposed to hydrogen every 6000 cycles (diamonds) and not exposed to hydrogen (triangles). Data for a control group of specimens that were intermittently cycled every 6000 cycles are also shown (squares). The control specimens (squares) were tested to compare with the intermittently hydrogen-exposed specimens (diamonds).

The data in figure 5 appears to show that hydrogen reduces the fatigue life of austenitic stainless steel fatigue samples in cycling bending. Table 2 given summary statistics for the data shown in figure 5.

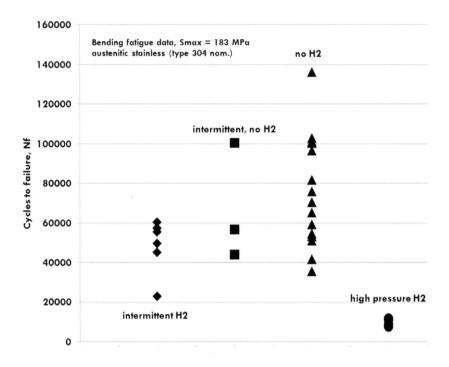

Figure 5. Cycles to failure for austenitic stainless specimens (type 304 nominally) cycled to failure with a maximum bending stress, S_{max}, of 183 MPa. Specimens were 0.9 mm thick. Specimens that were precharged with high pressure hydrogen (circles) show the lowest average cycles to failure.

Table 2.   Summary statistics to for austenitic specimens in bending (S_{max} = 183 MPa)

	intermittent 1atm H_2	intermittent cycling, no H_2	no H_2 exposure	high pressure H_2 precharge
n	6 specimens	3 specimens	15 specimens	5 specimens
avg N_f	48430 cycles	66970 cycles	75050 cycles	9460 cycles
stdev N_f	13660 cycles	29550 cycles	27840 cycles	1960 cycles

The effect of belt sanding the slag on the periphery of the fatigue samples is compared in Table 3. The data appears to indicate that no difference is detected by belt sanding the slag from the periphery of the fatigue sample before testing. The data shown in Table 3 are for samples that were not exposed to hydrogen.

Table 3. Summary statistics for bending fatigue samples, S_{max} = 183 MPa

material	austenitic stainless (type 304 nom.) 0.89 mm thickness (S_{max} = 183 MPa)	austenitic stainless (type 304 nom.) 0.89 mm thickness (S_{max} = 183 MPa)
sample prep	plasma cut only **no belt sand slag removal** *no hydrogen exposure*	plasma cut **belt sand slag removal** *no hydrogen exposure*
n	5 samples	10 samples
avg N_f	74020 cycles	75560 cycles
stdev N_f	27420 cycles	29510 cycles
avg d (mm)	12.8 mm	11.7 mm

Figure 6 shows maximum cyclic bending stress (S_{max}) as a function of the logarithm of cycles to failure (log N_f). Specimens that are highly precharged with hydrogen (circles) appear to have lower fatigue lives, for any applied bending stress. The intermittently cycled data are not shown. The data is shown as having no fatigue limit.

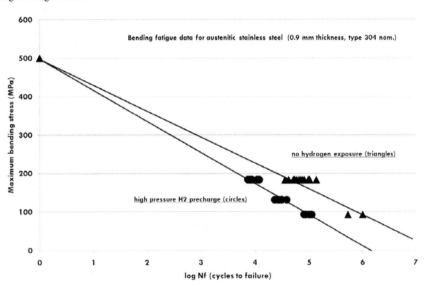

Figure 6. Maximum bending stress as a function of log cycles to failure for austenitic specimens. Specimens that are highly precharged with hydrogen (circles) appear to have lower fatigue lives, for any applied bending stress. The data is shown as having no fatigue limit.

DISCUSSION

Intermittent hydrogen cycling is an important protocol to develop because it allows for the testing of the effect of hydrogen when the equipment for in situ testing is not available. In the present work, samples were cycled 6000 times at a maximum bending stress of 183 MPa and then placed back into a 1 atm hydrogen environment for a minimum of 12 hours. The stress:cycles ratio is 183 MPa:6000 cycles, or 0.031 MPa cycles[-1]. Future intermittent cycling tests may strive for this stress:cycles ratio as a means of keeping the intermittent protocol reasonably consistent.

One of the advantages of intermittent cycling is that internal hydrogen embrittlement (IHE) is not a necessary possible failure mode, and therefore opens up the possibility of studying other types of materials that are embrittled through other means besides internal hydrogen. Because the hydrogen is applied intermittently, in a simulated continuous manner, hydrogen is always at an advancing crack tip. In situ hydrogen testing allows for the same capability, but the cost of creating in situ hydrogen testing set ups is higher than that for the intermittent set up.

The present investigation includes analytical and experimental work to determine the stress on the fatigue samples. The current method for calculating the maximum applied bending stress is by using elastic deflection equations for cantilevered beams. Because the profile of the fatigue specimens is tapered, the deflection equation estimations are only an approximation to the actual stress at the location of failure. An effort to model the state of stress in the specimens using analytical software has been initiated. Figure 7 shows typical output from a finite element calculation of stress in a sample at its maximum deflection. The software that was used for the analysis was SolidWorks. The specimen shown in fig. 7 appears to have a maximum bending stress of approximately 390 MPa at a maximum displacement of 4.5 mm. Future proposed work includes using strain gages on fixtured samples to measure the maximum strain at several deflections.

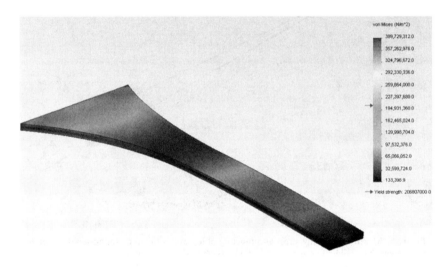

Fig. 7. Example of finite element analysis of a fatigue specimen at a maximum deflection of 4.5 mm. The maximum stress for the specimen modeled is 390 MPa.

CONCLUSIONS

Hydrogen was shown to reduce the bending fatigue life of austenitic samples of 0.9mm thickness. The most noticeable reduction in fatigue life was seen for samples that had been highly precharged with hydrogen. Intermittent exposure to 1 atm hydrogen may have an effect on reducing fatigue life but more data needs to be collected before a conclusion can be made.

The effect of belt sanding the periphery of fatigue samples did not appear to affect the fatigue life of plasma-cut austenitic samples. Samples are plasma-cut from austenitic stainless sheet to generate the required sample geometry, and the plasma leaves a light slag appearance on the periphery.

ACKNOWLEDGMENTS

The authors acknowledge John Cadwell, Larry Shockey, Rob Hardie, Patrick Nowacki and Beau Grillo of the Gonzaga University Engineering Department for their help with instrumentation, equipment assistance and materials preparation. The authors are grateful for hydrogen charging of specimens provided by Dr. Brian Somerday at Sandia National Laboratories.

REFERENCES

[1]Google Scholar search, May 23, 2011

[2]US DOE website, http://www1.eere.energy.gov/hydrogenandfuelcells/storage/hydrogen_storage.html, accessed May 24, 2011

[3]H.G. Nelson, "Hydrogen Embrittlement", Treatise on Materials Science and Technology, v. 25, Academic Press, ISBN 0-12-341825-9 (1983).

[4]C. San Marchi, B.P. Somerday, "Effects of High-pressure Gaseous Hydrogen on Structural Metals", SAE International, 2007-01-0433 (2006)

[5]C. San Marchi, B.P. Somerday, X. Tang, G.H. Schiroky, "Effects of Alloy Composition and Strain Hardening on Tensile Fracture of Hydrogen-precharged Type 316 Stainless Steels", Int'l J. Hydrogen Energy 33 (2008), 889-904.

[6]M. L. Martin, I.M. Robertson, P. Sofronis, nterpreting hydrogen-induced fracture surfaces in terms of deformation processes: A new approach, Acta Materialia, v. 59, issue 9, May 2011, pp. 3680-3687.

[7]P. Sofronis, Hydrogen Embrittlement of Pipelines: Fundamentals, Experiments, Modeling, DOE Project No. GO15045, period ending March 2010

[8]Y. Aoki, K. Kawamoto, Y. Oda, H. Noguchi, K. Higashida, "Fatigue Characteristics of a Type 304 Austenitic Stainless Steel in Hydrogen Gas Environment", International Journal of Fracture (2005) 133:277-288

[9]Y. Mine, K. Tachibana, Z. Horita, "Effect of High-Pressure Torsion Processing and Annealing on Hydrogen Embrittlement of Type 304 Metastable Austenitic Stainless Steel", Met. and Mat. Trans. A, v. 41A, December 2010.

[10]J. Zheng, X. Liu, P. Xu, P. Liu, Y. Zhao, J. Yang, "Development of High Pressure Gaseous Hydrogen Storage Technologies", International Journal of Hydrogen Energy, XXX (2011) 1-10.

[11]X.K. Sun, J. Xu, Y.Y. Li, Mater. Sci. Eng A114 (1989) 179-187.

[12]H.G. Nelson, J.E. Stein, "Gas Phase Hydrogen Permeation Through Alpha Iron, 4130 Steel and 304 Stainless Steel from less than 100°C to near 600°C" (NASA TN D-7265), Ames Research Center, National Aeronautics and Space Administration, Moffett Field CA (April 1973).

[13]W. Song, J. Du, Y. Xu, B. Long, J. Nuclear Mater. 246 (1997) 139-143.

[14]D.R. Begeal, J. Vac. Sci. Technol. 15 (1978) 1146-1154.

[15]P. Ferro, MS&T 2010 Ceramic Transactions, Houston TX (2010), John Wiley and Sons

[16]F.P. Beer, E.R. Johnston, J.T. DeWolf, Mechanics of Materials, 3rd ed., McGraw Hill Publishing, ISBN 0-07-365935-5 (2008)

LiMn$_x$Fe$_{1-x}$PO$_4$ GLASS AND GLASS-CERAMICS FOR LITHIUM ION BATTERY

Tsuyoshi Honma, Takayuki Komatsu

Department of Materials Science and Technology, Nagaoka University of Technology

Kamitomioka 1603-1, Nagaoka 940-2188, Japan

ABSTRACT

To clear the thermal property and electrical conductivity of LiMn$_x$Fe$_{1-x}$PO$_4$ glasses for the precursor in cathode materials for the lithium ion battery were examined. Glass formation region was being at x=0-0.8 in LiMn$_x$Fe$_{1-x}$PO$_4$ by melt quenching method in air. In LiFePO$_4$ (i.e. x=0) glass thermal stability for the crystallization is good compared with Mn rich glass. Electrical conductivity was decreased with increase of Mn content. The LiMn$_x$Fe$_{1-x}$PO$_4$/carbon composites were fabricated successfully using a glass powder crystallization processing, in which the mixture of glass powders and glucose was heat-treated at 560-800°C. The charge/discharge curves exhibit the plateaus at the voltage of ~3.4 and 4.1V for the Li metal.

INTRODUCTION

Lithium iron phosphate LiFePO$_4$ with an olivine structure has been proposed to be a potential candidate for use cathode materials for the next generation of rechargeable lithium ion batteries[1-2]. LiFePO$_4$ cathode materials have a high theoretical capacity of 170 mAh/g, are environmentally benign, thermal stable in the fully charged state, and have low raw materials costs. Numerous studies reported so far suggest that the key points to achieve high performances as cathode materials are to control or design particle size, morphology, and interface between LiFePO$_4$ crystal particles. LiFePO$_4$ is commonly synthesized via solid-state reactions, hydrothermal methods, and so on. Recently, we proposed new routes for the fabrication of phosphate-based lithium ion battery related materials such as olivine-type LiFePO$_4$, Nasicon-like Li$_3$Fe$_2$(PO$_4$)$_3$, and β-LiVOPO$_4$, in which the technique of glass crystallization was applied [3-8].

LiFePO$_4$ has a poor electronic and ion conductivity, resulting in the poor rate performance and thus limiting commercial applications. Many efforts have been made to improve the performance of LiFePO$_4$ cathode materials so far, including addition of conductive (Cu, Au, carbon) powders, doping of supervalence metal ions, carbon coating, and synthesis of nanoparticles. The iron (Fe^{2+}) site in LiFePO$_4$ can be substituted with other transition metal ions such as Mn^{2+}, i.e., LiMn$_x$Fe$_{1-x}$PO$_4$. In particular, it is noted that LiMnPO$_4$ having an olivine structure exhibits a plateau potential of 3.9 V in charge/discharge curves. This value of 3.9 V is higher than that (3.4 V) of LiFePO$_4$ [9,10].

Glass-ceramics method has the merit that it is possible to synthesize crystal phase from extremely homogeneous precursor grain, hence the reaction will be completed in a short time compared with a past technology. It is necessary to know vitrification and the crystallization behavior in the phosphate system that has been reported to apply the glass-ceramics technique in the field of lithium ion second batteries. The purpose of this study is to clear the thermal property and electrical conduction property of LiMn$_x$Fe$_{1-x}$PO$_4$ glasses and fabricate olivine-type LiMn$_x$Fe$_{1-x}$PO$_4$ crystals through the glass crystallization and to examine the lithium ion battery performance (electrochemical charge/discharge curves) for the glass-ceramics with LiMn$_x$Fe$_{1-x}$PO$_4$ crystals. In this study, we examined the glass formation and physical property (density and electronic conductivity) in the Li$_2$O-MnO$_2$-Fe$_2$O$_3$-P$_2$O$_5$ system, in particular for the stoichiometric compositions corresponding to LiMn$_x$Fe$_{1-x}$PO$_4$, i.e., the LiFePO$_4$-LiMnPO$_4$ pseudo binary system.

EXPERIMENTAL PROCEDURE

Glasses with the compositions of LiMn$_x$Fe$_{1-x}$PO$_4$ in the Li$_2$O-MnO$_2$-Fe$_2$O$_3$-P$_2$O$_5$ system were prepared using a conventional melt quenching method. Commercial powders of reagent grade LiPO$_3$, Fe$_2$O$_3$, MnO$_2$ were mixed well in a platinum crucible, and then the mixtures were melted at 1200°C in an electric furnace for 15 min in air. The melts were poured onto an iron plate and pressed to a thickness of 0.5 mm by another iron plate. The glass transition, T_g, and crystallization peak, T_p, temperatures were determined using differential thermal analysis (DTA) at a heating rate of 10 K/min. The electronic conductivity of glasses were measured by AC impedance method by using of impedance analyzer (HIOKI IM-3570). Gold electrodes (6mm ϕ) were prepared on both surface of glass plate with 0.3mm thickness by DC magnetron sputtering. AC impedance during 5Hz-1MHz was measured during temperature increasing from room temperature to 250°C.

Glass powders were obtained by crushing and grinding bulk plate glasses using a ball mill (Fritsch Co., premium line P-7). Glucose was added to glass powders in order to reduce the valences of Fe^{3+} to Fe^{2+} and Mn^{4+} to Mn^{2+} during crystallization. The amount of glucose was 5wt%. The mixtures of glass powders and glucose were heat treated at around temperatures crystallization peak temperature for 90 min in a reducing atmosphere of 7%H$_2$-93%Ar gas. The crystalline phase present in the crystallized samples was indentified by X-ray diffraction (XRD) analyses (CuKa radiation) at room temperature. The concentration of Fe^{2+} in the glass and glass-ceramics with the composition of LiFePO$_4$ was determined using a cerium redox titration method, in which 0.1N-Ce(SO$_4$)$_2$ solution as titrant and ortho-phenanthroline as indicator were used. The electrochemical charge/discharge curves were measured using the following procedures: Cathodes were prepared by mixing 80wt% glass-ceramics obtained (i.e.,active material), 15wt% graphitic carbon, and 5wt% polyvinyliden difluoride (PVDF) The mixture was then pressed into an Al thin sheet, and circular disks were prepared by cutting the sheet. Stainless test cells were constructed using a lithium metal anode and an electrolyte of 1M LiPF$_6$ consisting of a 1:1 solution of ethylene carbonate (EC) and diethyl carbonate (DEC). The lithium ion battery performance at room temperature was evaluated from charge/discharge measurements (Hokuto denko Co., HJ-1001). The cell potential was swept in the voltage range of 2.5 - 4.5 V, and the charge/discharge rate was kept as C/10 in each cell.

RESULTS AND DISCUSSION

PHYSICAL AND THERMAL PROPERTIES OF LiMn$_x$Fe$_{1-x}$PO$_4$ GLASSES

The DTA curves of bulk glasses measured on condition of 10 K/min is shown in Figure 1. And the value of the glass transition temperature T_g, the crystallization temperature T_c, and the scale of thermal stability against the crystallization ΔT ($= T_c - T_g$) in which were determined by DTA measurement, and the density in room temperature of LiMn$_x$Fe$_{1-x}$PO$_4$ glasses are also shown in TABLE I. LiFePO$_4$ (x= 0) glass showed a sharp crystallization peak at 553 °C, and ΔT was 71K. In previous study [4,5], it found that increase of ΔT is depends on the content of Fe^{3+} ions in glass, hence we are able to obtain relative thermal stable glass by melt-quenching method. The valence state in LiFePO$_4$ measured by chemical titration is as Fe^{3+}:Fe^{2+} = 87:13. In composition of x>0 which replaced by Mn in LiMn$_x$Fe$_{1-x}$PO$_4$ system, the crystallization peak changed broader and glass transition temperature and a crystallization peak shifted to lower temperature. As compared with that of x= 0, ΔT expands the value of that as 93K in the LiMn$_{0.2}$Fe$_{0.8}$FePO$_4$ (x= 0.2) glass. Therefore, it became clear that it is easier to vitrify by adding of Mn oxides. Although it is under investigation about the valence states of iron and manganese in detail, it is thought that coexistence of Fe^{2+}, Fe^{3+}, Mn^{2+}, and Mn^{3+} has contributed to the structural stability nature of phosphate glass.

Table I. Glass transition temperature T_g, crystallization temperature T_c, thermal stability of glass ΔT from DTA measurement and density of quenched bulk glass ρ in LiMn$_x$Fe$_{1-x}$PO$_4$ glasses

x	T_g (°C)	T_c (°C)	ΔT (K)	ρ (g/cm^3)
0	497	553	71	3.198
0.2	471	540	93	3.291
0.4	468	534	96	3.264
0.5	465	537	97	3.269
0.6	450	520	97	3.270
0.8	436	505	91	3.301

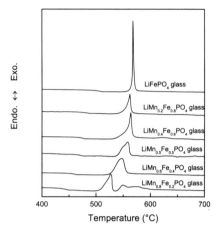

Figure 1. DTA patterns for the LiMn$_x$Fe$_{1-x}$PO$_4$ bulk glasses. Heating rate was fixed as 10K/min.

The compositional dependence of the density of the rapid-cooling glass in room temperature is shown in Figure 2.

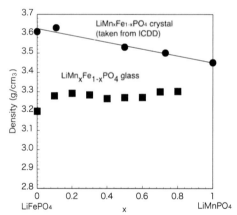

Figure 2. Bulk density at room temperature as a function of Mn content in LiMn$_x$Fe$_{1-x}$PO$_4$ glasses. Theoretical density calculated from ICDD data is also shown.

The theoretical density of the corresponding crystal calculated from the ICDD database is also shown. In the system of LiMn$_x$Fe$_{1-x}$PO$_4$, LiFePO$_4$ glass has the lowest density with 3.198 gcm^3, and density became higher with increase of the amount of Mn substitution. This relation is contrary to that for the theoretical density in corresponding crystal. It is considered that glass structure approaching the crystal structure in Mn rich composition that vitrification becomes difficult.

ELECTRICAL CONDUCTIVITY OF LiMn$_x$Fe$_{1-x}$PO$_4$ GLASS

The temperature dependence of the electrical conductivity measured by AC impedance method in LiFePO$_4$ glass is shown in Fig. 3. Activation energy for the conduction was calculated as 0.50 eV by using Arrhenius equation. Since this value is the almost same value in LiFePO$_4$ crystal, it seems that a conduction mechanism is the same. Olivine type LiFePO$_4$ shows azimuthal dependence in Li$^+$ ion conduction, and favorable conduction directions are along with b- and c-axis [11]. The conductivity of the single crystal was also shown in Fig. 3, glass showed the intermediate of that of LiFePO$_4$ single crystal.

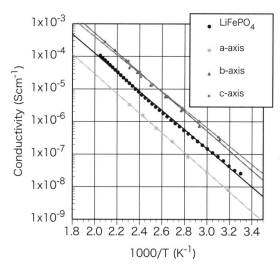

Figure 3. Electronic conductivity of LiFePO$_4$ glass plate measured by AC impedance method as a function of the temperature.

The Arrhenius plots in electronic conductivity of Mn substituted LiMn$_x$Fe$_{1-x}$PO$_4$ glasses are shown in Fig. 4. The compositional dependence of the activation energy for the conduction calculated from the

Arrhenius plot is shown in Fig. 5.

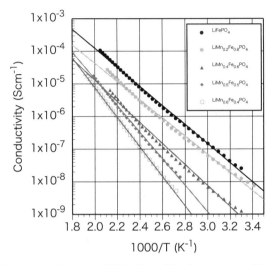

Figure 4. Electronic conductivity of LiMnxFe1-xPO4 glasses measured by AC impedance method as a function of the temperature.

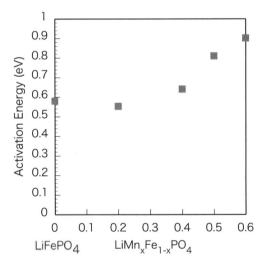

Figure 5. Activation energy for the conductivity in LiMn$_x$Fe$_{1-x}$PO$_4$ glasses.

In a series of LiMn$_x$Fe$_{1-x}$PO$_4$ glass, Glass structure becomes dense with increase of Mn substitution, involving with the reduction of free volume, hence the conductivity fell due to the decrease of Li$^+$ ion diffusion. The conductivity fell remarkably with increase of the amount of Mn oxide substitution, and activation energy also increased. The conductivity at room temperature of LiMnPO$_4$ crystal is about 10^{-11}Scm^{-1} which is much less than that of LiFePO$_4$ ($\sim$10^{-8}Scm^{-1}), and low conductivity affect to battery performance.

ELECTROCHEMICAL PROPERTIES OF LiMnxFe1-xPO4 GLASS CERAMICS

The Raman scattering spectrum of LiFePO$_4$ glass and crystallized glass powder is shown in Figure 6. In LiFePO$_4$ glass-ceramics powder strong and sharp peak is being at 955 cm^{-1} corresponding to the presence of isolated (PO$_4$)$^{3-}$ unit. On the other hand the broad Raman band is confirmed around 1022 cm^{-1} in LiFePO$_4$ glass. Raman bands around 1000cm^{-1} is attributed to the vibration of pyrophosphate (P$_2$O$_7$)$^{4-}$ units. It is clear that the phosphate framework differ with a crystal and glass, respectively. It is considered that the difference in phosphate framework brought low density and increase of free volume in glass.

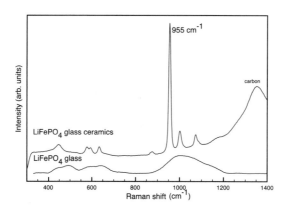

Figure 6. Raman scattering spectra of the as quenched LiFePO$_4$ glass and glass-ceramics powder.

As shown in Figure 7 the electrochemical charge and discharge curves for the cells consisting of LiMn$_x$Fe$_{1-x}$PO$_4$ glass-ceramics as the cathode and lithium metal as the anode. The charging curve for LiMn$_{0.5}$Fe$_{0.5}$PO$_4$ glass-ceramics shows a clear plateau at 3.4 and 4.1 V, which represents red/ox potentials of Fe^{3+}/Fe^{2+} and Mn^{3+}/Mn^{2+}. In other LiMn$_x$Fe$_{1-x}$PO$_4$ glass-ceramics, a clear plateau is not observed at 4.1 V. The electrochemical capacity decreases with increasing Mn content. The low conductivity reflects that it is necessary to fabricate much small particle in Mn rich glass-ceramics. Further studies on the carbon coating and morphology of LiMn$_x$Fe$_{1-x}$PO$_4$ crystals prepared by the glass crystallization method would be necessary to improve lithium ion battery performances.

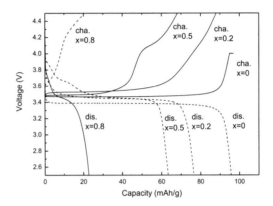

Figure 7. Electrochemical charge and discharge curves for the cells consisting of LiMn$_x$Fe$_{1-x}$PO$_4$ glass-ceramics as the cathode and lithium metal as the anode. The cell potential was swept in the voltage range of 2.5 - 4.5 V, and the charge/discharge rate was kept as C/10 in each cell.

CONCLUSION

Glass formation and electronic conductivity of LiMn$_x$Fe$_{1-x}$PO$_4$ glasses are examined. Glass formation stability increased with Mn content during x=0.2~0.5. Although the Li$^+$ conducting pass disturbed by random structure in LiFePO$_4$ precursor glass however ionic conductivity was exhibit intermediate value between a, c axis and b-axis in LiFePO$_4$ single crystal. It seems low density of LiFePO$_4$ glass making the free volume to assist the ionic conduction. By substituting with Mn ionic conductivity was drastically decreased, in which reflects the increasing of glass density. By means of Glass-ceramics technique single-phase olivine phase which working as cathode materials in Li+ ion battery are obtained in whole compositions. Glass ceramics technique is also useful for the fabrication of active materials that has extremely homogenous composition in Li$^+$ battery without any byproducts.

ACKNOWLEDGEMENTS

This work was supported from the Grant-in-Aid for Scientific Research from the Ministry of Education, Science, Sports, Culture, and Technology, Japan and Program for Developing the Supporting System for Global Multidisciplinary Engineering Establishment, Nagaoka University of Technology, Japan.

REFERENCES

[1] A.K. Padhi, K.S. Nanjundaswamy, J.B. Goodenough, Phospho-olivines as positive-electrode materials for rechargeable lithium batteries, *J. Electrochem. Soc.*, **144**, 1188-1194 (1997).

[2] A.K. Padhi, K.S. Nanjundaswamy, C. Masquelier, S. Okada, J.B. Goodenough, Effect of structure on the Fe3+/Fe2+ redox couple in iron phosphates, *J. Electrochem. Soc.* **144**, 1609-1613 (1997).

[3] K. Hirose, T. Honma, Y. Benino, T. Komatsu, Glass-ceramics with LiFePO$_4$ crystals and oriented crystal line patterning in glass by YAG laser irradiation, *Solid State Ionics*, **178**, 801-807 (2007).

[4] K. Hirose, T. Honma, Y. Doi, Y. Hinatsu, T. Komatsu, Moessbauer analysis of Fe ion state in lithium iron phosphate glasses and their glass-ceramics with olivine-type LiFePO4 crystals, *Solid State Commun.* **146**, 273-277 (2008).

[5] T. Honma, K. Hirose, T. Komatsu, T. Sato, S. Marukane, Fabrication of LiFePO$_4$/carbon composites by glass powder crystallization processing and their battery performance, *J. Non-Cryst. Solids*, **356**, 3032-3036 (2010).

[6] T. Honma, K. Nagamine, T. Komatsu, Fabrication of olivine-type LiMn$_x$Fe$_{1-x}$PO$_4$ crystals via the glass-ceramic route and their lithium ion battery performance, *Ceramics International*, **36**, 1137-1141 (2010).

[7] Nagamine, K. Hirose, T. Honma, T. Komatsu, Lithium ion conductive glass-ceramics with Li$_3$Fe$_2$(PO$_4$)$_3$ and YAG laser-induced local crystallization in lithium iron phosphate glasses, Solid State Ionics, **179**, 508-515 (2008).

[8] K. Nagamine, T. Honma, T. Komatsu, Selective synthesis of lithium ion conductive β-LiVOPO4 crystals via glass-ceramic processing, J. Am. Ceram. Soc. 91 (2008) 3920-3925.

[9] C. M. Burba, and R. Frech, Local structure in the Li-ion battery cathode material Li$_x$(Mn$_y$Fe$_{1-y}$)PO$_4$ for 0<x≤1 and y=0.0, 0.5 and 1.0, *J. Power Sources*, **172**, 870-876 (2007) .

[10] A. Yamada, M. Hosoyo, S. Chung, Y. Kudo, K. Hinokuma, K. Liu, Y. Nishi, Olivine-type cathodes: Achievements and problems, *J. Power Sources*, **119-121**, 232-238 (2003).

[11] J.Y. Li, W.L. Yao, S. Martin, D. Vaknin, Lithium ion conductivity in single crystal LiFePO$_4$, *Solid State Ionics*, **179**, 2016 (2008).

THE ABSORPTION OF HYDROGEN ON LOW PRESSURE HYDRIDE MATERIALS

Gregg A. Morgan, Jr. and Paul S. Korinko
Savannah River National Laboratory
Aiken, SC 29808

ABSTRACT

For this study, hydrogen getter materials (Zircaloy-4 and pure zirconium) that have a high affinity for hydrogen (and low overpressure) have been investigated to determine the hydrogen equilibrium pressure on Zircaloy-4 and pure zirconium. These materials, as with most getter materials, offered significant challenges to overcome given the low hydrogen equilibrium pressure for the temperature range of interest. Hydrogen-zirconium data exists for pure zirconium at 500°C and the corresponding hydrogen overpressure is roughly 0.01 torr. This manuscript presents the results of the equilibrium pressures for the absorption and desorption of hydrogen on zirconium materials at temperatures ranging from 400°C to 600°C. The equilibrium pressures in this temperature region range from 150 mtorr at 600°C to less than 0.1 mtorr at 400°C. It has been shown that the Zircaloy-4 and zirconium samples are extremely prone to surface oxidation prior to and during heating. This oxidation precludes the hydrogen uptake, and therefore samples must be heated under a minimum vacuum of 5×10^{-6} torr. In addition, the Zircaloy-4 samples should be heated at a sufficiently low rate to maintain the system pressure below 0.5 mtorr since an increase in pressure above 0.5 mtorr could possibly hinder the H_2 absorption kinetics due to surface contamination. The results of this study and the details of the testing protocol will be discussed.

INTRODUCTION

The capture, storage, and release of hydrogen and hydrogen isotopes are important features for safe implementation of the hydrogen economy. Sample preparation methods, testing methods, and data on the behavior of hydrogen getter materials (metals and intermetallic compounds) are of particular interest for additional research and development. Zirconium and its alloys are important materials for technological applications in the field of energy production, in particular as hydrogen storage materials[1], as well as in the nuclear industry as hydrogen getter materials[2]. Given that the equilibrium pressure of zirconium is < 0.1 mtorr at 400°C, zirconium and its alloys are well suited for getter materials instead of reversible hydrogen storage materials. Zirconium can be loaded to a Q/M (where Q is the moles of H, D, or T and M is the moles of metal atoms) to 2.0, which corresponds to a weight percent of roughly 2.2%. The reaction of hydrogen with Zr at its surface, e.g. chemisorption or transport of hydrogen atoms from the surface to the bulk determines the H_2 absorption, which is of importance in H_2 storage applications. The zirconium-hydrogen system has been extensively studied by a variety of techniques over the years[1-10]. However, very little information is currently available regarding the equilibrium pressures of the H_2 isotopes adsorbed on zirconium or its alloys at low temperatures (i.e. 300°C – 500°C).

Hydrogen gettering materials (Zircaloy-4 and pure zirconium) with exceptionally high affinity for hydrogen (and low overpressure) have recently been investigated to determine the hydrogen equilibrium pressure on Zircaloy-4 and pure zirconium. The primary objective of this work was to determine the equilibrium pressure-composition-temperature data for the absorption and desorption of protium and deuterium on Zircaloy-4 and pure zirconium. Protium-zirconium hydride data exist for temperatures in excess of 500°C, and at a temperature of 500°C correspond to a pressure of roughly 10 mtorr[11]. The equilibrium pressure for the absorption and desorption of protium on Zircaloy-4 at temperatures ranging from 400°C to 600°C have been determined. The absorption equilibrium pressures in this temperature region range from 150 mtorr at 600°C to less than 0.1 mtorr at 400°C. As a result of this study, it has been determined that the Zircaloy-4 and zirconium samples are extremely prone to surface oxidation, which inhibits the hydrogen uptake, and therefore must be heated under a minimum vacuum of 5×10^{-6} torr. In addition, the Zircaloy-4 samples should be heated at a rate of 5°C per minute in order to maintain the system pressure below 0.5 mtorr. An increase in pressure above 0.5 mtorr will hinder the H_2 absorption kinetics due to surface contamination.

EXPERIMENTAL

Two different gas handling manifolds were used to complete the experiments reported here, Manifold I and Manifold II. The two manifolds are shown in Figure 1 and Figure 2, respectively. Both manifolds consist of a series of valves, stainless steel tubing, calibrated volumes, and pressure transducers. Both gas handling manifolds are evacuated and maintained at the high vacuum range (1×10^{-7} torr). Manifold I is equipped with an Alcatel Drytel 31 vacuum pump and Manifold II is equipped with a Pfeiffer turbomolecular pump back by a Varian tri-scroll vacuum pump.

Figure 1. (Color Online) Photograph of Manifold I, showing the sample cell, the heater assembly, the 10K torr MKS Baratron, the 1 torr MKS Baratron.

Manifold I is equipped with a 0 – 1 Torr MKS Baratron and a 0 – 10K Torr MKS Baratron, which were both calibrated along with the associated displays. Manifold II is equipped with a 0 – 100 mtorr MKS Baratron for low pressure measurements. The internal volume of both manifolds was approximately 130 mL, excluding the volume of the sample vessels, which was roughly 15mL. Manifold I was primarily constructed of gasketed fittings, whereas Manifold II was constructed primarily of specially cleaned, welded fittings with gaskets only at the valves and sample ports. The minimization of gasketed fittings would ideally result in a system that was able to achieve a lower ultimate pressure. The initial results were obtained on Manifold I with a 0-1 Torr MKS Baratron, which was only accurate down to 0.1 mtorr. Manifold II was subsequently designed to include a 100 mtorr Baratron, which would be accurate down to 0.01 mtorr.

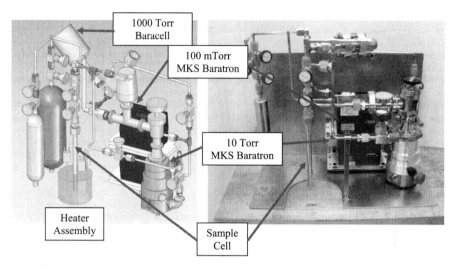

Figure 2. (Color Online) Solid Edge model of Manifold II (left) and photograph of the as constructed Manifold II (right).

The sample vessels were constructed of Type 316 stainless steel (SS) 0.5" x 0.049" tube with a Type 316 SS endcap welded to the tube and a ½" Type 316 L VCR fitting welded to the other end of the vessel. After construction, the sample vessels were thoroughly degassed in a large vacuum furnace. The vessels were loaded into the furnace and then evacuated (<1 × 10⁻⁵ torr at room temperature). The furnace was heated at 15°C/minute to 1050°C and held for four hours while continually being pumped. The furnace was then cooled under vacuum, and the sample vessels removed, capped, and placed in a bag that was purged with I$_2$ until they were ready for use. This procedure removed (or minimized) any potential impurities present in the stainless steel sample vessel that would desorb when the vessel was initially heated for testing. Due to the low hydrogen equilibrium pressure of zirconium at the target temperatures[11], it was expected that any impurities would adversely affect the overall performance of the zirconium getter. In addition, any off-gassing from the sample vessels would adversely influence the background pressure in the test manifold.

The sample vessels could be reused for new samples if the necessary measures were taken to clean the vessels of adsorbed hydrogen. The empty vessel was heated to 600°C for at least 12 hours under high vacuum conditions to desorb any hydrogen that may have diffused into the stainless steel. The heater assembly consisted of a temperature controller, an over temperature controller, and a split cylindrical Watlow ceramic heater. The two halves of the heater were placed around the sample vessel. The sample vessel was placed completely inside the heater assembly in order to heat the entire vessel. The sample vessel was cooled under vacuum then backfilled with 1 atm of argon prior to opening the sample vessel to load a sample.

Samples consisted of one inch long Zircaloy-4 tubes and pure zirconium strips that had been weighed to achieve nominal 1 gram samples. The samples were mechanically abraded with P1200 grit grinding paper on all surfaces to remove any surface oxide that may have been present since preliminary testing had demonstrated that a surface oxide layer would inhibit the hydriding ability of the Zircaloy-4. The H$_2$ absorption was considerably slower on samples that were not cleaned, either mechanically or chemically, due to the presence of the oxide film. The samples were rinsed in acetone and in 200 proof ethanol, wiped with lint free wipes and cotton swabs and allowed to air dry before being weighed on an analytical balance. The offgassed, nitrogen or argon filled sample vessel was opened and the new sample loaded such that the sample vessel was open to the atmosphere for a minimum amount of time, typically less than 20 seconds.

The sample vessel was attached to the manifold with a 0.5 micron filter gasket and the system was evacuated overnight. Prior to heating the sample, the system with an internal volume of ~150 mL passed a rate of rise of less than 0.05 mtorr in 10 minutes. The bottom two inches of the sample vessel were heated incrementally (in 100°C steps) to the desired temperature (typically 550°C) while being evacuated. Following each increase in temperature of the sample vessel, the system was evacuated until P < 0.10 mtorr. The maximum temperature (550°C) was held for several hours (3-4 hours) such that the rate of rise was less than 0.05 mtorr in 10 minutes. The middle of the sample vessel was roughly 180°C and the top of the sample vessel was roughly 80°C when the sample was at 600°C.

High purity hydrogen (Air Liquide, Research grade, 99.9995 %) and high purity deuterium (Spectra gases, Research Grade, 99.999 %) were used. Two different methods of testing were utilized to measure the equilibrium pressure of the H_2 on the zirconium and Zircaloy-4. In the first method, Method A, the samples were evacuated overnight, tested for rate of rise at room temperature, step heated to the loading temperature (550°C), tested for rate of rise, evacuated, loaded with protium or deuterium to the target Q/Zr (hydrogen atom/zirconium atom), then absorption tested by cooling in roughly 30°C increments to temperatures as low as 300°C. For Method B, the samples were evacuated overnight, tested for rate of rise at room temperature, step heated to the desired loading temperature, rate of rise tested, loaded with H_2 to the desired Q/Zr, then cooled to room temperature, and the over pressure evacuated. Desorption tested proceeded by heating the sample to 300°C initially followed by roughly 50°C increments to 550°C. The sample was allowed to reach equilibrium and data was taken at each step. Absorption testing followed after the sample temperature reached 550°C by cooling in 30°C increments to as low as 300°C; data was collected in the same manner. The Method B technique was developed because of concerns of He ingrowth associated with proposed subsequent tritium testing and its effect on measured pressures.[12]

RESULTS AND DISCUSSION
Manifold I: Zircaloy-4

Zr-4 alloy samples were loaded with either protium or deuterium to Q/Zr of 0.5 on nominally 1 gram pieces. Method A testing for absorption was used to collect the equilibrium pressure data. The results of the testing are listed in Table I and displayed graphically in Figure 3. Note that the data is not linear below about 450°C. The loss of linearity is attributed to a combination of off-gassing, in-leakage, and gas impurities. One can see from these data that the lower temperature data are clearly not on the same slope as the higher temperature data. There is also some off-set between the literature data[11] and the experimental data, since no repeat data were generated; it is unknown where this difference arises. There is also a pressure difference between the protium and deuterium, with deuterium having a higher measured presure. The original plot of the referenced literature data is shown in Figure 4, while the interpolated and extrapolated data is listed in Table II.

Table I: Equilibrium Absorption Pressures of Protium and Deuterium over Zr-4 using Manifold I, Method A

Equilibrium Pressures for Protium		Equilibrium Pressures for Deuterium	
Temperature	Pressure	Temperature	Pressure
(°C)	(mtorr)	(°C)	(mtorr)
599.10	87.78	598.80	137.03
548.90	14.64	548.84	20.85
498.50	2.28	498.48	3.01
448.50	0.35	448.71	0.41
398.50	0.10	397.64	0.08
348.30	0.08	347.65	0.08

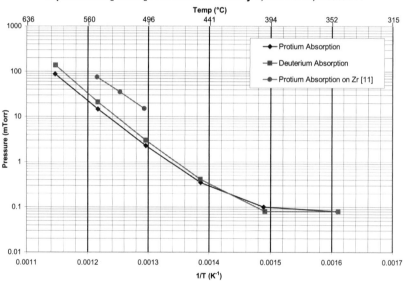

Figure 3. (Color Online) Graphical representation of the equilibrium pressure data for the absorption of H_2 on Zircaloy-4 using Manifold I, Method A for a loading of Q/Zr of 0.5. The literature data is from Reference 11.

Table II: Equilibrium Pressure of Protium over Zirconium from Mueller [11].

	Equilibrium Pressures on Zr for Protium		
	Pressure (mtorr)		
Q/M	500°C	525°C	550°C
0.20	12.3	32.7	68
0.50	12.7	33	68
0.80	12.9	33	68
1.10	13	33	80
1.40	30	71	105

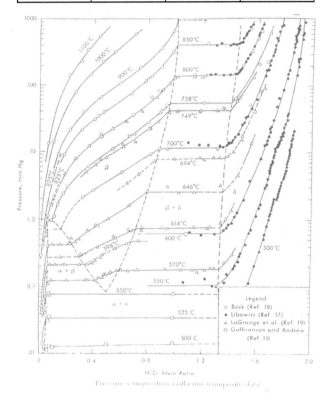

Figure 4. Reference data for equilibrium pressure of H_2 on zirconium. Taken from Ref. 11.

Manifold II: Zircaloy-4

Manifold II was used to determine the equilibrium pressures of H_2 absorbed on Zircaloy-4 with both Method A and Method B. The data are presented in Table III and Table IV, respectively, and graphically in Figure 5. It is apparent from the data that values above about 400°C yield linear data with increasing temperature. This result is consistent with the results obtained from the testing completed on Manifold I, which showed similar behavior and a significant deviation from linearity at the lower temperatures, which can be attributed to inleakage and offgassing.

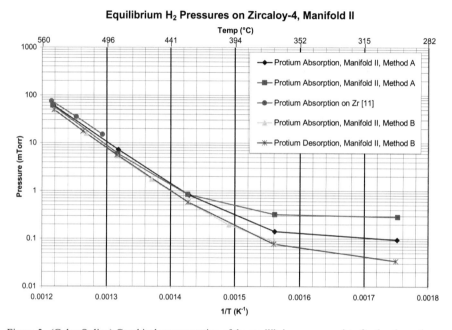

Figure 5. (Color Online) Graphical representation of the equilibrium pressure data for the absorption of H_2 on Zircaloy-4 using Manifold II, Methods A and B for a loading of H/Zr of 0.5.

From these results, it can be determined that Manifold II yields linear data down to ~375°C, which is ~75°C lower than where the data from Method A deviates from linearity. The equilibrium pressures measured using Method B are also lower than the equilibrium pressures measured using Method A. At 550°C, the measured equilibrium pressure using Method B was roughly 19% lower than similar data using Method A and at 425°C the data measured using Method B was 28% lower than the corresponding data point using Method A. It is also interesting to note that there is very little hysteresis between the desorption and absorption data when Method B is used. The absence of hysteresis between the absorption and desorption of hydrogen can possibly be a characteristic of a low pressure hydride material. There is very little literature evidence on the isothermal absorption and desorption on getter type hydride materials and no evidence to support or refute the claim regarding the absence hysteresis. It is also possible that there could be some hysteresis between the hydrogen absorption and desorption data and that it is within the noise of the pressure measurements and is therefore not detectable.

Table III: Equilibrium pressure for the Absorption of H_2 on Zr-4 Using Manifold II, Method A with a loading of 0.5 H/Zr.

Equilibrium Pressures for Protium Manifold II		Equilibrium Pressures for Protium Manifold II (Re-test)	
Temperature	Pressure	Temperature	Pressure
(°C)	(mtorr)	(°C)	(mtorr)
548.08	61.611	548.94	60.253
485.74	7.227	486.12	5.931
426.73	0.808	427.76	0.839
367.18	0.141	367.17	0.322
297.85	0.095	297.70	0.286

Table IV: Equilibrium pressure for the Absorption of H_2 on Zr-4 Using Manifold II, Method B with a loading of 0.5 H/Zr.

Zircaloy-4 H/Zr = 0.5, Method B					
Temperature	Pressure	Comment	Temperature	Pressure	Comment
(°C)	(mtorr)		(°C)	(mtorr)	
550.48	62.778	Initial Loading	547.40	49.801	Common Point
298.56	0.034	Desorption	515.40	15.767	Absorption
367.91	0.077	Desorption	486.55	5.351	Absorption
427.54	0.581	Desorption	456.78	1.763	Absorption
486.94	5.644	Desorption	427.36	0.569	Absorption
518.54	17.474	Desorption	397.71	0.196	Absorption
547.40	49.801	Desorption	367.26	0.086	Absorption

Manifold II: Pure Zirconium

Pure zirconium strip samples were loaded to a protium to zirconium ratio of 0.5. The equilibrium pressures were determined using Method B. The measured equilibrium pressures are presented in Figure 6 and Table V. These data indicate very little hysteresis between the absorption and desorption data. The data are also slightly lower than the literature values.

Table V: Test results from loading Zr strip to an H/Zr ratio of 0.5 with results from Manifold II, Method B.

Pure Zr H/Zr = 0.5			Pure Zr H/Zr = 0.5		
Temperature	Pressure	Comment	Temperature	Pressure	Comment
(°C)	(mtorr)		(°C)	(mtorr)	
547.83	58.145	Initial Loading	547.01	55.404	Common Point
298.49	0.041	Desorption	516.72	18.661	Absorption
370.03	0.082	Desorption	487.10	6.256	Absorption
398.01	0.196	Desorption	457.70	1.995	Absorption
427.95	0.622	Desorption	427.17	0.623	Absorption
457.64	1.994	Desorption	399.00	0.202	Absorption
487.21	6.193	Desorption	368.30	0.078	Absorption
516.96	18.269	Desorption	332.74	0.042	Absorption
547.01	55.404	Desorption	297.99	0.036	Absorption

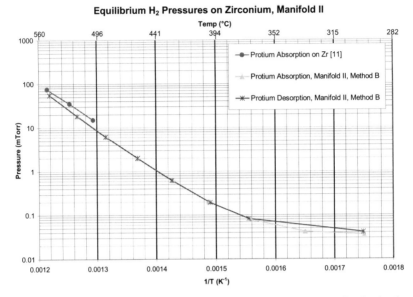

Figure 6. (Color Online) Graphical representation of the equilibrium pressure data for the absorption of H_2 on zirconium using Manifold II, Method B for a loading of H/Zr of 0.5.

SUMMARY OF ALL RESULTS
All of the data that has been collected on Zircaloy-4 and zirconium using Manifold I and II (Methods A and B) is graphically displayed in Figure 7. In all cases the equilibrium pressures measured on Zircaloy-4 and pure zirconium are lower than the previous measurements reported by Meuller.[11] At least two possibilities exist, one is that the alloying effects on the Zircaloy-4 could suppress the pressures and the second is that the new data represents an improvement in the test technique with better manifold materials and instrumentation. The data collected on Manifold II is much closer to the literature values reported by Mueller[11] than the data collected on Manifold I.

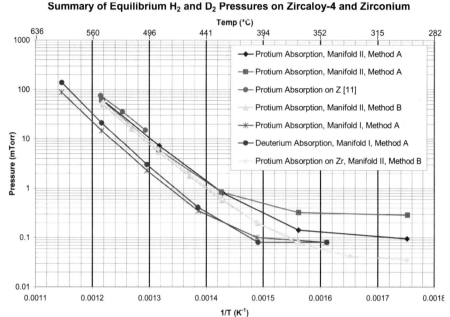

Figure 7. (Color Online) Summary of equilibrium pressure for H_2 and D_2 Absorption on Zircaloy-4 and Zirconium using Manifolds A and B.

The lower equilibrium pressure measured on Manifold I can possibly be attributed to less off-gassing and potential contamination from the manifold. Manifold I has been in service for a significantly longer period of time than Manifold II and has been evacuated for much of that time. The continual evacuation of Manifold I likely has resulted in a reduction in the level of off-gassing and impurities coming from the walls of the manifold. Manifold II was constructed of cleaned stainless steel for this work and was also vacuum degassed in a furnace prior to assembly. Even though these precautions were taken, it is still possible that the various components of Manifold II are the source of small amount of off-gassing and contamination, resulting in a slightly higher equilibrium pressure.

The zirconium surface is very prone to surface oxidation regardless of the type of oxidant (oxygen, water, water vapor, CO, etc.)[13]. Due to the diffusivity of oxygen in zirconium[14,15] and due to the chemical affinity of Zr for oxygen, the formation of the oxide occurs immediately upon contact even at the lowest pressures of the oxidant. Given this information it is imperative to evacuate the zirconium and Zircaloy-4 samples as soon as possible following the mechanical abrading. The pressure in the sample cell should be as low as possible prior to heating the sample to the activation temperature. Ideally a pressure $<5\times10^{-6}$ torr is necessary before heating the sample to the activation temperature. Additionally, it has been determined that a step-wise heating of the sample may be necessary (depending on the pumping speed of the system) to maintain the pressure in the system below 0.5 mtorr.

SUMMARY AND CONCLUSIONS

A manifold was designed, built, proof-tested, and utilized to obtain equilibrium pressure data over Zircaloy-4 and pure zirconium. These getter materials offered significant challenges to overcome given the low hydrogen equilibrium pressure in the temperature range of 400°C to 600°C. Previously zirconium hydride data was available down to only 500°C and results are reported here for temperatures down to 400°C. The H_2 equilibrium pressure on Zircaloy-4 was measured to be 0.1 mtorr at 400°C. In order to obtain equilibrium pressures for hydrogen on zirconium it was necessary to minimize the surface oxidation by evacuating the sample to $<5\times10^{-6}$ torr prior to heating the sample. In addition it was also necessary to minimize the pressure increase in the system during the heating of the sample. This can be accomplished by heating the sample incrementally and allowing the pressure to return to the base level before increasing the temperature.

ACKNOWLEDGMENTS

The authors gratefully acknowledge Jody Dye and James Klein of Savannah River National Laboratory, Donna Hasty of the Savannah River Site, and Dave Senor of Pacific Northwest National Laboratory for their support of this work. The manuscript has been prepared for the U.S. Department of Energy under Contract Number DE-AC09-08SR22470.

1.   K. Ojima and K. Ueda, *App. Surf. Sci.*, **165**, 149 (2000).
2.   C.S. Zhang, B. Li, and P. R. Norton, *Surf. Sci.*, **346**, 206 (1996).
3.   J. A. Llouger and G. N. Walton, *J. Nuc. Mats.*, **97**, 185 (1981).
4.   M. Yamamoto, S. Naito, M. Mabuchi, and T. Hashino, *J. Phys. Chem.*, **96**, 3409 (1992).
5.   C. S. Zhang, B. J. Flinn, K. Griffiths, and P. R. Norton, *J. Vac. Sci. Tech. A*,, **10**, 2560 (1992).
6.   G. R. Corallo, D. A. Asbury, R. E. Gilbert, and G. B. Hoflund, *Langmuir*, **4**, 158 (1988).
7.   G. G. Libowitz, J. Nuc. Mats. **5**, 228 (1962).
8.   R. D. Penzhorn, M. Devillers, and M. Sirch, *J. Nuc. Mats.*, **179-181**, 863 (1991).
9.   Y. Naik, G. A. Rama Rao, and V. Venugopal, *Intermetallics*, **9**, 309 (2001).
10.   K. A. Terrani, M. Balooch, S. Wongsawaeng, S. Jaiyen, and D. R. Olander, *J. Nuc. Mats.*, **397**, 61 (2010).
11.   W. M. Mueller, J. P. Blackledge, G. G. Libowitz, Metal Hydrides, Academic Press, New York, 1968.
12.   G. C. Staack and J. E. Klein, *Fus. Sci. Tech.,* **60**, 1479 (2011).
13.   R. A. Causey, D. F. Cowgill, and R. H. Nilson, "Review of the Oxidation Rate of Zirconium Alloys," Sandia Report, SAND2005-6006 (2005).
14.   A. Grandjean and Y. Serruys, *J. Nuc. Mats.*, **273**, 111 (1999).
15.   R. A. Perkins, *J. Nuc. Mats.*, **68**, 148 (1977).

POLYMETHYLATED PHENANTHRENES AS A LIQUID MEDIA FOR HYDROGEN STORAGE

Dr. Mikhail Redko
Powdermet, Inc.
24112 Rockwell Drive, Euclid, Ohio, 44117, USA

ABSTRACT

A mixture of polymethylphenanthrenes was studies as a prospective material for hydrogen storage in a liquid carrier that, hypothetically, may store more than 6% hydrogen by weight. The mixture was synthesized by non-selective reductive phenanthrene methylation with Na/MeI. Subsequent hydrogenation of that liquid resulted in a mixture of saturated polycyclic hydrocarbons that remained liquid down to -25°C.

INTRODUCTION

One of the major factors preventing hydrogen from being used as a transportation fuel is the lack of materials and/or equipment capable of storing it in significant amounts. As an example, hydrogen constitutes 3 wt % of the commercial steel cylinders where it is kept at 200atm. Correspondingly, it would be advantageous to find and/or develop a system with higher hydrogen content. If that hydrogen carrier turns out to be a liquid that would allow employing the modern infrastructure that distributes liquid hydrocarbon fuel (gasoline, kerosene) which would, in turn, eliminate the potential expenses related to building of an alternative infrastructure necessary to handle compressed or liquefied hydrogen. It would also allow refueling the hydrogen tank on a timescale of few minutes, which is significantly lower than what it would have taken if hydrogen was stored in form of solid hydrides that were yet to be made onboard in the course of a chemical reaction. Unlike liquid hydrogen that creates a low temperature hazard and requires continuous energy input for refrigeration, an organic liquid can be stored for an extended period of time without the need to be continuously refrigerated. Also, unlike pressurized hydrogen, it does not present a high-pressure hazard.

Liquid aromatic hydrocarbons have long been considered as potential hydrogen storing materials. Their hydrogenation results in saturated cyclic hydrocarbons, which may provide hydrogen upon catalytic thermal dehydrogenation onboard vehicles (1):

$$C_xH_y + zH_2 \rightleftarrows C_xH_{y+2z} \qquad (1)$$

Methylcyclohexane-toluene[1,2] and decalin/tetralin-naphthalene[3-5] transformations have long been studied for hydrogen storage. While the second set of substances includes naphthalene, which is solid at room temperature (m.p. 80°C), these and related processes share the same disadvantage, namely, the necessity to supply too much energy to drive the endothermic dehydrogenation process.

The energy required to dehydrogenate the lowest alkanes is comparable to energy released from hydrogen oxidation ($\Delta H^f_{H2O,l}$ = -285 kJ/mol; $\Delta H^f_{H2O,g}$ = -242 kJ/mol). For example, ethane dehydrogenation enthalpy (2) is 136 kJ/mol, i.e., $\sim$-50% of the water formation enthalpy. If the energy released upon hydrogen oxidation is spent on ethane dehydrogenation with $\sim$50% efficiency, the corresponding device would be spending all of the hydrogen oxidation energy for the ethane dehydrogenation process and there would be no energy left to perform useful work.

$$C_2H_6 \rightarrow C_2H_4 + H_2 \qquad (2)$$

The high dehydrogenation enthalpy is the reason why methylcyclohexane will unlikely find an application as a hydrogen storage material: its specific energy density, that takes into account the energy required for the dehydrogenation, is 1.83kW*h/kg (if hydrogen is subsequently oxidized into liquid water) or 1.49kW*h/kg (if hydrogen is oxidized into water vapor), so that both values are below the DOE 2010 target value of 2kW*h/kg.

An ideal hydrogen carrier should have an equilibrium hydrogen pressure of 1atm at ambient temperature. Given that the standard enthalpy of hydrogen $S°_{H2}$=130 J/K*mol, and assuming that the total entropy change in the dehydrogenation reaction (1) equals the one of hydrogen formation, the equilibrium pressure of 1atm will be achieved if the dehydrogenation enthalpy equals
ΔH = 130 J/K*mol*298K ~40 kJ/mol H_2.

In search for an optimal hydrogen storage media, an analysis of dehydrogenation enthalpies of representative hydrocarbons has been performed using the NIST database. It demonstrated that the dehydrogenation enthalpy was becoming less and less positive as substitution of the alkanes was increasing. An additional stabilization of the newly formed unsaturated compounds originates from aromatization of the unsaturated compounds. The dehydrogenation enthalpy of 9,10-dihydrophenanthrene (50 kJ/mol) approaches that "ideal" ΔH, while, in the limiting case (graphene/graphane transformation), the computed dehydrogenation enthalpy becomes so low (21.6 kJ/mol H_2) that it renders the corresponding hydrocarbon (graphane) thermodynamically unstable.[6] This indicated that the hydrogenated counterparts of the aromatic compounds containing more condensed rings than phenanthrene or having substituents on the phenanthrene rings would be characterized by dehydrogenation enthalpy being even closer to 40 kJ/mol.

The energy to drive the dehydrogenation process may come from the oxidation of the hydrogen carrier itself. Thus, in one version, the catalytic total oxidation of toluene was used to bring the required calories to perform hydrogen generation from methylcyclohexane.[7] The obvious drawback of this process was the oxidative destruction of the hydrogen carrier (toluene). In a somewhat different version of the process, fluorene oxidation into fluorenone was utilized to drive the dehydrogenation of perhydrofluorene into fluorene.[8] Fluorenone was subsequently deoxyhydrogenated and hydrogenated back into the perhydroflyorene. An application of this cycle would allow use of perhydrophenanthrene as a mobile hydrogen source; the fluorenone would be subsequently hydrogenated in stationary devices. However, just like with naphthalene-based hydrogen storage system, the high melting points of fluorene (117°C) and fluorenone (85°C) will complicate the vehicle fueling and unloading. A more advanced set of similar aromatic, hydrogenated and oxidized compounds would have to have all of the components be liquids at ambient conditions.

The high melting points of polyaromatic hydrocarbons (PAHs), resulting from efficient packing of their molecules, unavoidably complicate their application for hydrogen storage. Some of those values are presented in Table I; they show that even the compounds containing only two annulated aromatic rings are solids at room temperature; as the number of the annulated rings increases, the melting points increase even more.

Table I. Melting Points of Representative Aromatic Compounds and Their Methylated Derivatives

Compound	Melting point, °C	Compound	Melting point, °C
Benzene	5	Methylbenzene (toluene)	-93
Naphthalene	80	1-Methylnaphthalene	-22
Biphenyl	70	2-Methylnaphthalene	35
Fluorene	116	9-Phenanthrene	90
Fluorenone	83.5	2-Phenanthrene	56
Phenanthrene	101	1-Methylanthracene	86
Anthracene	218	9-Methylanthracene	81

Alkyl-substituted PAHs often have melting points significantly lower than those of the parent compounds, as illustrated in Table I. Since the methyl group is not going to give off hydrogen in the process of dehydrogenation, it will be ballast, diluting the hydrogen-absorbing media and decreasing the energy content of the liquid. Therefore, to increase the media hydrogen content, the number of methyl groups on aromatic molecules will have to be minimized. Also, only two methylarenes have melting points below ambient temperature, so the methylation alone is not a sufficient factor for the liquefaction of the hydrogen storage media.

Another factor affecting the melting point of a substance is its purity: mixtures of chemically non-interacting substances that do not co-crystallize have lower melting points than the individual components. Therefore, to ensure that the hydrogen storage media remains liquid at ambient temperature, one can use a mixture (perhaps eutectic) of methylaromatic compounds, rather than a single substance.

An additional factor restricting the number of compounds suitable for hydrogen storage is that the boiling point should be higher than the dehydrogenation temperature (typically ~300°C) to avoid energy wastes associated with the carrier's phase transitions, like evaporation of methylcyclohexane (b.p. 101°C; $\Delta_{vap}H° = 35$ kJ/mol).

Based on these premises, we propose a mixture of methylphenanthrenes (9-methylphenanthrene, b.p. 353°C) as a novel liquid hydrogen storage media.

A traditional way for the synthesis of methylarenes is via parent chloromethylation followed by reduction of chloromethylarene into methylarene:

$$Ar - H \xrightarrow{HCl, CH_2O} Ar - CH_2Cl \xrightarrow{H_2 / Pd} Ar - CH_3 \qquad (3)$$

According to literature sources, chloromethylation of phenanthene under optimized conditions[9,10] yielded mostly 9-chloromethylphenanthrene. References on the subsequent dechlorination step have not been identified in the literature, but a related process, debenzylation, is well documented.[11, 12, 13] Those reactions are typically performed by hydrogenation with molecular hydrogen at 1atm or hydrogen donors (HCOOH, NH$_4$HCOO, N$_2$H$_4$, NaH$_2$PO$_2$, etc.) in presence of Pd/C catalyst. The goal of this Project, however, was to generate a mixture of isomeric compounds (chloromethylphenanthrenes) with comparable proportions, rather than maximize the yield of one particular isomer. Correspondingly, a number of attempts were made to adjust the conditions of the chloromethylating procedure so as to generate the desired mixture. The sequence of chloromethylation followed by reduction did result in a mixture of methylphenanthrenes that was liquid at room temperature, but the yield was too low, so an alternative synthetic procedure was developed.

Reductive methylation of phenanthrene (Fig 1a→b) turned out to give practically acceptable yields. It was performed by reduction of phenanthrene with sodium metal, or NaK alloy, in THF, to highly colored phenanthrene radical-anion salts M$_n$C$_{14}$H$_{10}$ (M=Na, K; n=1,2) followed by their methylation with methyl iodide and quenching with proton sources. The resulting mixture of polymethylated and partially hydrogenated phenanthrene derivatives did not freeze, even at T=-25°C, indicating the success of the initial plan. Dehydrogenation of those compounds with elemental sulfur (Fig 1b→c) allowed differentiating between aromatic protons and protons from the methyl groups so that the average degree of methylation could be determined by the integration of lines in the ^{1}H NMR spectrum. Hydrogenation of the methylated party hydrogenated mixture (Fig 1b→d) was performed at 150°C, 700 Psi H$_2$ in a high-pressure Parr reactor in presence of 5% Ru/C catalyst. The hydrogenated mixture (Fig 1d), which is currently considered as a potential hydrogen source, remained liquid at T=-25°C as well.

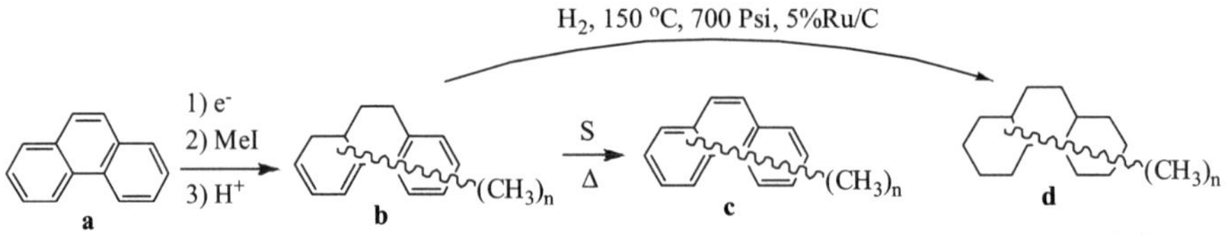

Figure 1: (a) phenanthrene; (b) mixture of (poly)methylated, partly reduced phenanthrenes; (c) mixture of methylphenanthrenes; (d) mixture of methylated hydrogenated phenanthrenes.

EXPERIMENTAL

Materials and Equipment

Anthacene, phenanthrene, pyrene and perylene were purchased from TCI, trioxane from Aldrich, paraform and $ZnCl_2$ from Baker, Na from ACP Chemicals Inc., iodomethane from EMD Chemicals Inc., K from Alfa Aesar. The chemicals were used as received. GCMS of the products of phenanthrene chloromethylation were performed on Hewlett Packard 5890 Series II Gas Chromatograph coupled to Trio-1 mass spectrometer. GCMS of the phenanthrene reductive methylation products were performed using Agilent Technologies 6890 GC coupled to Agilent Technologies 5973 MS; injection was performed by Agilent Technologies 7863 injector in the Michigan State University mass-spectroscopic facilities. CH analysis was performed by Micro-Analysis, Inc. Hydrogenation was performed in a 300mL SS Parr reactor.

Melting of Anthracene/Phenanthrene/Pyrene/Perylene Mixture

A mixture of anthacene (m.p. 218°C), phenanthrene (m.p. 98°C), pyrene (m.p. 148°C) and perylene (m.p. 280°C), 0.5g each, has been prepared in a test tube and heated up on an oil bath at a rate of 1°C/min until the whole mixture liquefied. After that, it was allowed to cool down. The melting/solidification points were recorded and their averages reported.

Phenanthrene Chloromethylation

In an exemplary experiment, phenanthrene (0.5g, 2.81 mmoles), trioxane (0.5g, 5.55 mmoles or 16.7 mmoles CH_2O equivalents), and 36% aq. HCl (5mL, 5.9 mmoles) were refluxed in 10 mL acetic acid in oil bath for 5 hours. After that, the mixture was diluted with chloroform (10 mL), the chloroform layer was rinsed with saturated aqueous $(NH_4)_2CO_3$ solution to neutralize the acids and then evaporated to viscous liquid. That liquid was redissolved in $CDCl_3$ and analyzed by ^1H NMR and GCMS.

Reduction of (Chloromethyl)Phenanthrenes

The mixture from the preceding experiment (mostly chloromethylphenanthrenes), 5% Pd/C catalyst (100 mg), ammonium formate (1.0 g) and methanol (50 mL) was placed into a 100 mL round bottom flask. The content of the flask was refluxed overnight, cooled down and filtered. Sizable chunks of yellow organic solids, insoluble in methanol, supposedly polymeric materials, were found mixed with the catalyst. The methanol solution was filtered and evaporated in nitrogen stream to give mostly white crystalline precipitate of unreacted NH_4HCOO. That crude product was stirred with chloroform to dissolve the desired aromatic hydrocarbons, after that chloroform was filtered again and evaporated to yield 48 mg of a liquid mixture of methylphenanthrenes with unreacted phenanthrene, as evidenced by ^1H NMR and GCMS (9% counting the whole mixture as methylphenanthrenes).

Phenanthrene Reductive Methylation

Phenanthrene (Alfa Aesar, 90%, 100g) and Na metal (63g, cut into 16 pieces) were placed into a 3L round bottom flask containing a magnet bar. THF dried by contact with Na metal (1L) was added so the mixture soon becomes dark-green. The solution was stirred for two hours under nitrogen, then mixture of 100mL methyl iodide with 100mL THF was added in five portions with eight hours breaks between additions while cooling the flask with ice during those additions. Those breaks were taken to ensure that the aromatic compounds would react with Na and convert back into anion-radicals. After all of the methyl iodide was added, the mixture was allowed to stand for eight more hours so that the newly formed anion-radicals destroy the excess of MeI, then it was quenched with 30mL t-BuOH and then 100mL H_2O. The solvent (mostly THF) was evaporated in a stream of nitrogen, then the remaining organic liquid was re-dissolved in hexane (800mL), filtered through Celite into a 1L flask and then the hexane was evaporated in a stream of nitrogen to result in a yellow oily liquid. Yield: 117.5g. Phase transition behavior of the resulting mixture was investigated by keeping the reaction mixture for 24 hours at –25 °C in a freezer.

Determination of Methylation Degree

Mixture of reduced methylated phenanthrenes (0.6 g) was heated with elemental sulfur (0.6g) to 230°C in a 5mL flask and the evolved hydrogen sulfide collected. Subsequently, a product sample was dissolved in $CDCl_3$ and the degree of methylation deduced from the ratio of areas from CH_3 vs. C_{Aryl}-H protons in the ^1H NMR spectrum.

Hydrogenation of Methylated Phenanthrenes

Mixture of methylated phenanthrenes obtained by reductive methylation was diluted to 100mL with cyclohexane and loaded into a 300mL Parr reactor with 0.32g 50% wet 5%Ru/C catalyst. The reactor was purged, then pressurized to 800Psi with hydrogen and heated upon stirring. To identify the lowest temperature at which the hydrogenation reaction takes place at an acceptable rate, the temperature was ramped (80°C, 120°C, 150°C). Samples were periodically taken out of the reactor and dried in stream of nitrogen to evaporate cyclohexane. The degree of hydrogenation was estimated from the ratio of aromatic to aliphatic peak areas in ^1H NMR spectra. Subsequently, the content of the reactor was filtered into a 100mL flask, the cyclohexane was evaporated in a nitrogen stream and the flask was placed into a freezer (-25°C) to check if the hydrogenated mixture of methylphenanthrenes would freeze.

Flammability and Density of the Hydrogenated Mixture

An attempt was made to ignite a vapor of the hydrogenated mixture with open flame at 80 °C, which did not result in inflammation. To measure the liquid density, 1.000 cm^3 of the liquid (measured by volumetric pipette) was weighed.

RESULTS AND DISCUSSION

The eutectic point of the anthacene/phenanthrene/pyrene/perylene mixture was found to be 90°C; all of the solids melted at 180°C. Even though the decrease in melting point, resulting from mixing, has been demonstrated, the eutectic temperature of that mixture was still too high for the hydrogen storage application. That demonstrated the chemical modifications of the aromatic compounds outlined in the original Proposal, such as methylation, were indeed necessary.

In the phenanthrene chloromethylation experiments, significant amounts of new viscous organic liquid, heavier than the acetic acid, typically formed. That liquid completely dissolved in chloroform. Its GC was dominated by two broad bands, consistent with formation of two major classes of similar types of products. The M/z ratio of the light fraction band, illustrated in Figure 2, was consistent with the presence of parent phenanthrene, chloromethyl- and (acetoxymethyl)phenanthrene or multiple isomers of those compounds. The M/z=368 peak of the heavy fraction (Figure 3) was consistent with the formation of bis(phenanthrenyl)methane indicating the onset of polycondensation processes. ^{1}H NMR spectra of the reaction mixture products (not shown) typically had a number of sharp peaks in the 4.5-5.5 ppm region, assigned to various isomers of Ar-CH$_2$-X (X=Cl, OH, O$_2$CCH$_3$, etc.) and Ar-CH$_2$-Ar'. Peaks with δ=2.1-2.2 ppm were ascribed to residual acetic acid and its derivatives.

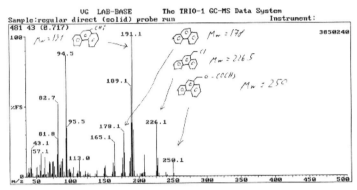

Figure 2. Mass spectrum of the light fraction of phenanthrene chloromethylation product mixture.

Performing the reaction at longer times (24 hours instead of five hours) typically resulted in hardening of that product whose ^{1}H NMR spectrum, instead of distinct individual lines, exhibited broad bumps typical for the polymers. No bis(chloromethyl)phenanthrene isomers have ever been detected among the reaction products by either NMR or GCMS.

The NMR of the product's mixture resulting from dechlorination of chloromethylphenanthrenes (highly soluble in methanol) had two strong sharp peaks at 2.734 ppm and 2.761 ppm, ascribed to methyl protons of two of the predominant methylphenanthrene isomers. Two more smaller, but still prominent, singlets in the 2.5-2.7 ppm region were ascribed to the minor isomers. Several small peaks in the 4.6-5.2 ppm region were ascribed to X-CH$_2$-X, C$_{14}$H$_9$-CH$_2$-X (X=OH, Cl, OAc, other electro withdrawing groups) derivatives that seemed to avoid the reduction.

The prominent peak at 4.897 ppm was ascribed to C$_{14}$H$_9$-CH$_2$-C$_{14}$H$_9$ (calculated δ=4.84 ppm). In contrast, the ^{1}H NMR spectrum of the methanol-insoluble highly viscous products had significant area under the peaks in the 4.6-5.2 ppm region (with the most prominent singlet at 4.897 ppm) and barely visible peaks in the 2.5-2.8 ppm region, indicating that it mostly consisted of the polycondensed oligomers that had low solubility in methanol.

Gas chromatograph of the methanol-soluble product mixture (Fig 4) demonstrates the predominance of one product with T$_{ret}$=17.58 sec (M/z=192), ascribed to the major methylphenanthrene isomer, most likely 9-methylphenanthrene. A minor peak with T$_{ret}$ =16.88 sec (M/z=192) was ascribed to a minor isomer. A prominent peak at T$_{ret}$=15.82 sec (M/z=178) was ascribed to the parent phenanthrene. The peaks that could have been assigned to dimethylphenanthrenes were insignificantly small.

The low yield of the desired product, presence of unreacted phenanthrene as well as polycondensation products in the final mixture illustrated that the rate constant of the reaction between phenanthrene and (chloromethyl)phenanthrene was higher than that between phenanthrene and chloromethylating agent(s) such as $^+$CH$_2$OH. As a result of that, the desired (chloromethyl)phenanthrenes tended to convert into polycondensation products, rather than accumulate in the mixture. Thus, the mixture of methylphenanthrenes for the hydrogen absorption has been synthesized and it did turn out to be a liquid at room temperature. However, the low yield of the desired product indicated that the synthetic procedure had to be modified.

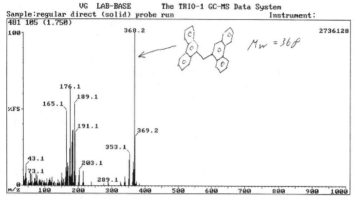

Figure 3. Mass spectrum of heavy fraction of phenanthrene chloromethylation product mixture.

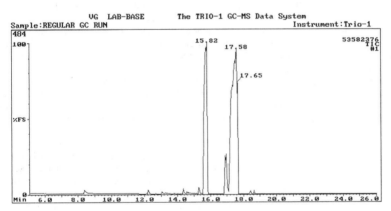

Figure 4. GC of the MeOH-soluble fraction of (chloromethyl)phenanthene reduction.

A number of experiments were performed to change the reaction conditions so as to increase the yields of the chloromethylphenanthrenes and methylphenanthrenes resulting from them. In those experiments the chloromethylation was performed using aqueous HCl-acetic acid mixtures of different compositions or sulfolane. A $ZnCl_2$ catalyst was sometimes employed; trioxane or paraform were used as sources of formaldehyde, $POCl_3$ was used sometimes as an additional source of HCl and dehydrating agent and source of HCl. The time varied from one hour (typically no reaction took place at that timescale) to three days; temperature interval typically ranged from room temperature (~20°C) to ~120°C (boiling point of acetic acid-phosphoric acid mixtures). Those reactions typically resulted either in low conversion (when time was short and/or reaction time was too low) or formation of viscous, insoluble, highly colored products of phenanthrene-formaldehyde polycondensation. Thus, this route has achieved no significant increase of the methylphenanthrene yield, so an alternative procedure (reductive methylation) had to be developed.

Reductive methylation was performed by stirring the THF phenanthrene solution with alkali metal (NaK alloy or just Na) for approximately five hours to maximize the reduction of phenanthrene into anions and anion radicals. Methyl iodide was added dropwise upon cooling the whole flask with ice until the solution had almost completely lost its color. A white solid (NaI) suspended in the solution could be seen at this point. Then the MeI addition was stopped and the mixture was stirred to ensure the reduction of phenanthrene and (poly)methylphenanthrenes into their anion radicals until it got dark again. A gas evolution was notable during methyl iodide addition. The ultimate yield of the liquid product was higher than the calculated amount, most probably, because of the solvent left in the product.

The ^{1}H NMR and GCMS of the phenanthrene reductive methylation product mixture indicated that phenanthrene had indeed been converted into a wide array of methylated partly hydrogenated aromatic hydrocarbons containing up to five methyl groups per phenanthrene backbone. Up to 40 individual peaks have been resolved in the chromatogram, but the actual number of products, for sure, exceeded that amount. No solids were formed in the mixture for 24 hours at −25°C, thus illustrating the melting point depression in a mixture of substances.

Heating that mixture of hydrocarbons with sulfur led to gas evolution at T=190-210 °C. The process halted abruptly and no gas evolved when the temperature was further increased to 230°C indicating the completion of dehydrogenation. The volume of the evolved gas (105 cm^3) corresponded to ~4.3 mmols of H$_2$S. The ^{1}H NMR spectrum of the product had a ratio H_{Ar}:$H_{Ar\text{-}CH3}$= 10:2, corresponding to the degree of alkylation of ~0.6 CH$_3$ groups per aromatic molecule even though 2.7MeI/phenanthrene ratio was used in the synthesis. Phenanthrene peaks were clearly seen in the aromatic part of the spectrum. These results and gas evolved during methyl iodide addition to the solution point at the probable ethane formation that apparently competed with reductive methylation process and decreased the yield of the desired product.

After addition of the first methyl group the reactivity of the phenanthrene derivatives went down, so the subsequent addition of the methyl groups happened at significantly lower rate, even when excess of metal and methyl iodide were used. The presence of detectable amounts of pentamethylphenanthrenes indicated that, unlike in the chloromethylation process, the reductive methylation still proceeded far beyond the attachment of the first methyl group for some molecules.

The average molecular weight of the mixture is (178+0.6*14)g/mol=186.4 g/mol, which means that 0.6g/(186.4g/mol) ~ 3.22*10^{-3}mols of methylphenanthrenes were dehydrogenated. The known amount of hydrogen sulfide evolved allows writing down the average formula of the mixture components as C$_{14}$H$_{10}$(CH$_3$)$_{0.6}$H$_{2.07}$. The fact that the number of extra hydrogen atoms attached to phenanthrene molecule exceeded the number of methyl groups indicated that phenanthrene underwent a Birch-type reduction concurring with the methylation step by reaction with sodium metal and t-butanol.

During the studies on hydrogenation of methylphenanthrenes, the temperature of the reactor was increased stepwise so as to identify the temperature at which the hydrogenation does proceed to the desired degree, but at the same time avoid decomposition of the organics on the catalyst. The process was too slow at T<150°C, so that 24 hours at 80°C and 24 hours at 120°C were still insufficient for the hydrogenation. Thus, 150°C was considered as minimal temperature for the hydrogenation. The ratio of aryl/alkyl hydrogens in the ^{1}H NMR spectrum was decreasing as temperature and hydrogenation time progressed. That ratio was found to be equal 11/203 in the final spectrum; no cyclohexane peak could be discerned in the aliphatic part, so that the error caused by its presence was considered negligible. Given that ratio and that the aromatic molecules in the starting material had an average 0.6 methyl groups per phenanthrene skeleton, that corresponds to presence of ~13% of the aromatic rings in the final material that avoided hydrogenation. Elemental composition was C-87.67%, H-11.13% (total – 98.80%), thus indicating a presence of impurity (most likely water). Assuming that there were in average 0.6 CH_3 groups per phenanthrene skeleton and that all of the carbon and hydrogen belonged to the hydrogenated compounds, this showed that an average (polymethyl)phenanthrene molecule combined with 5.29 H_2 molecules or 76% of the maximal possible number ($7H_2$). The discrepancy in the results can be attributed to the impurities leftover from handling the reagents via different procedures and likely presence of $R^1R^2C=CR^3R^4$ double bonds in the final mixture that would affect the elemental composition, but would not be identified by ^1H NMR.

Given the random distribution of the methyl groups among the reactants, presence of substances with up to five methyl groups per phenanthrene system (as shown by GCMS) and difficulty in hydrogenation of the highly substituted rings, it is likely that the highly methylated aromatic rings avoided hydrogenation. Correspondingly, the possible ways to increase the degree of hydrogenation might include the increase of the hydrogenation time, temperature and pressure – or to adjust the mixture composition so that it does not have substances with more than one methyl group per phenanthrene skeleton.

The hydrogenated mixture did not freeze after 24 hours of standing at -25°C which indicated that the liquid hydrogen source developed within this project is unlikely to freeze even at adverse weather conditions. Its density was found to be 0.97±0.02g/cm^3. Given that every polymethylphenanthrene molecule can combine with up to $7H_2$ molecules ($\Delta M_W=14g/mol$) and that the experimentally found degree of hydrogenation equals 100%-13%=87%, that allows one to calculate the hydrogen and energy contents of the liquid hydrogen carrier synthesized within this project. These enthalpy calculations were performed for the 87% hydrogenated material, 100% hydrogenated material (calculated on the assumption that upon further development the hydrogenation degree will be made quantitative), with water formed in both gaseous and liquid states. The results are presented in Table II; they indicate that the hydrogen weight content in that material exceeds the 2011 DOE Interim hydrogen storage requirements for the year of 2017.[16] It needs to be pointed out, though, that the evolution of more than 6% wt hydrogen is yet to be demonstrated. Also, the DOE targets refer to the whole system while the data on hydrogen content in methylphenanthrenes are related to the material only. Taking the rest of the storage system into account will decrease the hydrogen system content, but the excess of the calculated material hydrogen content over the system required value shows that the DOE system value may still be achievable.

The high flash point of the liquid (<80 °C) is an additional safety factor characterizing the newly developed hydrogen storage media.

NIST database was used to find or estimate the thermodynamic functions of the substances involved and changes of those functions during chemical reactions pertinent to the new hydrogen storage liquid. Since the corresponding data were unavailable for the methylphenanthrene derivatives, data on phenanthrene or isomeric anthracene derivatives were used instead. Assuming that oxidation of methylphenanthrenes goes to methylphenanthrenequinones, and does not involve oxidation of methyl groups, and that the enthalpies of the corresponding reactions are equal to those of parent phenanthrene, a cycle illustrated in Figure 5 has been proposed.

Practically, we expect that the heat necessary for the release of 7 moles of H_2 in the dehydrogenation step at ~ 200 °C will be provided by aerial catalytic oxidation of methylphenanthrenes into methylphenanthrenequinone. The carrier recharging requires 10 moles of hydrogen so the efficiency of hydrogen storage in this scheme is 70%. The other 30% can be considered as an energetic "storage fee" analogous to energy spent on hydrogen compression, liquefaction, and/or other transformations performed in alternative hydrogen storage schemes to win over the entropy factor that counters the reduction of volume of hydrogen. A comparison with a similar scheme using fluorene and its derivatives[8] for hydrogen storage demonstrates that the fluorene-based system can release hydrogen in almost thermoneutral manner, where the enthalpy required for the dehydrogenation of perhydrofluorene is provided by fluorene oxidation into fluorenone. That factor makes the flourene-based scheme somewhat more hydrogen efficient: it utilizes $6H_2$ out of $8H_2$, i.e., 75%.

Table II. Hydrogen and Energy Content in Hydrogenated Aromatics

System	Specific energy* kWh/kg	Specific energy** kWh/kg	Energy density* kWh/L	Energy density** kWh/L	Stored hydrogen mass %	Stored hydrogen density, g/L
DOE 2017 Interim Storage targets	1.8	1.8	1.3	1.3	5.5	40
87% hydrogenated methylphenanthrenes	2.43	2.06	2.36	2.00	6.13	58.4
100% hydrogenated methylphenanthrenes	2.77	2.35	2.69	2.28	6.99	67.8

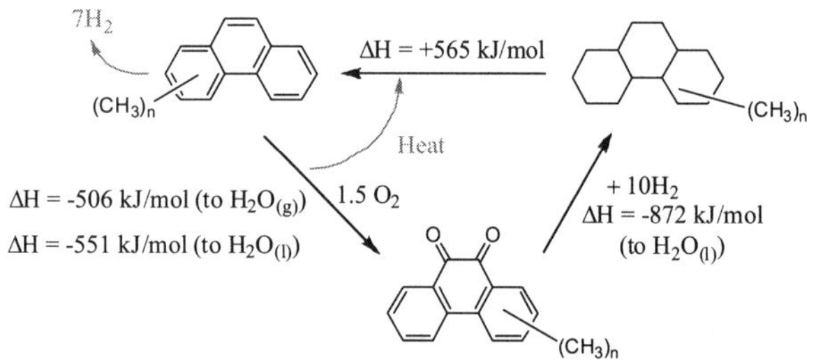

Figure 5. Proposed hydrogen storage in methylphenanthrenes

Progressive methylation of aromatic rings decreases the dehydrogenation enthalpy thanks to the Van-der-Waals repulsion between axial methyl groups and other cyclohexane ring substituents. For example, the dehydrogenation enthalpies of cyclohexane, methylcyclohexane and 1,4-(cis)-dimethylcyclohexane to benzene, toluene and p-xylene (all liquids) decrease from 206.7 to 202.2 to 191.2 kJ/mol. It is expected, therefore, that the dehydrogenation enthalpy of methylphenanthrenes will be lower than 565 kJ/mol shown in Figure 5, so the methylphenanthrene scheme may become fully authothermal. That will bring the overall energetics of the methylphenanthrene scheme in par with the fluorene-based cycle. The additional advantage of the proposed scheme is that the melting points of its components will be depressed in comparison with fluorene (m.p. 117°C) and fluorenone (m.p. 85°C).

The current project research directions include demonstration and optimization of the remaining steps of the hydrogen storage cycle illustrated in Figure 5, namely, thermal catalytic dehydrogenation of the mixture of perhydropolymethylphenanthrenes into polymethylphenanthrenes, oxidation of that mixture into polymethylphenanthrenequinones and construction of a reactor utilizing the oxidation heat to drive the dehydrogenation step.

FOOTNOTES

*Assuming that water from the hydrogen oxidation process is obtained in a liquid state.
**Assuming that water from the hydrogen oxidation process is obtained in a gaseous state.

CONCLUSIONS

Mixture of partially hydrogenated methylphenanthrenes has been prepared by reductive alkylation of phenanthrene. That product was further hydrogenated into a mixture of perhydropolymethylphenanthrenes and the calculated content of potentially releasable hydrogen exceeds 6wt%. The media is expected to release hydrogen as a result of an autothermal process. The main advantage over the similar process that utilizes fluorene as a storage media is that all of the cycle components will stay in a liquid state at all conceivable storage, refueling and operation temperatures. The experimental determination of the amount of releasable hydrogen and other material properties are currently under investigation.

ACKNOWLEDGMENTS

I would like to acknowledge Professor James E. Jackson (MSU) for the organizational help and fruitful scientific discussions, Philip Atem for his help in phenanthrene chloromethylation experiments and the MSU Biochemistry Mass Spectroscopic Facilities for performing GCMS analyses. The experimental work was performed under AF contract # **FA8501-09-P-0231**.

REFERENCES

1. M.Usman, F. Alhumaidan, D. Cresswell, A. Garforth, Methylcyclohexane dehydrogenation - A convenient way for hydrogen storage. *AIChE Annual Meeting, Conference Proceedings, Jashville, TJ, United States, Jov. 8-13,* 2009, Pages usman1/1-usman1/7, Conference, Computer Optical Disk (2009).
2. A. Shukla, P. Gosavi, J. Pande, V. Kumar, K. Chary, R. Biniwale, Efficient hydrogen supply through catalytic dehydrogenation of methylcyclohexane over Pt/metal oxide catalysts. *International Journal of Hydrogen Energy,* **35**(9), 4020-4026 (2010).
3. A. Shono, T. Hashimoto, S. Hodoshima, K. Satoh, Y. Saito, Continuous catalytic dehydrogenation of decalin under mild conditions. *Journal of Chemical Engineering of Japan*, **39**(2), 211-215 (2006).
4. Y. Wang, N. Shah, F. Huggins, G. Huffman, Hydrogen Production by Catalytic Dehydrogenation of Tetralin and Decalin Over Stacked Cone Carbon Nanotube-Supported Pt Catalysts. *Energy & Fuels*, **20**(6), 2612-2615 (2006).
5. B. Wang, W. Goodman, G. Froment, Kinetic modeling of pure hydrogen production from decalin. *Journal of Catalysis*, 253(2), 229-238 (2008).
6. G. Joelson, Denille B. de Lima, L. Assali, and J. Justo. Group IV Graphene- and Graphane-Like Nanosheets. *J. Phys. Chem. C*, *115* (27), 13242–13246 (2011).
7. P. Kerleau, Y. Swesi, V. Meille, I. Pitault, F. Heurtaux, Total catalytic oxidation of a side-product for an autothermal restoring hydrogen process. *Catalysis Today,* 157(1-4), 321-326 (2010).
8. A. Cooper et al; IV.B.2. Hydrogen Storage by Reversible Hydrogenation of Liquid-Phase Hydrogen Carriers. *Air Products and Chemicals, Inc., DOE Contract # DE-FC36-04GO14009.*
9. F. Fernandez et al. A useful access to 9-phenanthrylmethyl derivatives. *Synthesis*, 10, 802-3 (1988).
10. W. Pfeiffer, Synthesis by substitution of carbonyl oxygen. *Science of Synthesis*, Volume Date 2006, 35, 155-165 (2007).
11. Y. Dai, S. Dong, Z. Pan, J. Chen, Research on and application of Pd/C catalysts for catalytic hydrogenolysis debenzylation. *Gongye Cuihua*, 19(4), 7-10 (2011).
12. H. Donkervoort, Blocking group removal catalysts for green technologies. *Speciality Chemicals Magazine,* 28(3), 32-33 (2008).
13. K. Bujnowski, A. Adamczyk, L. Synoradzki, o-Aminomethyl derivatives of phenols. Part 1. Benzylamines: properties, structure, synthesis and purification. *Organic Preparations and Procedures International,* 39(2), 153-184 (2007).

14. M. Chono, Dehydrogenation and oxidative dehydrogenation catalysts. 2. Processes for dehydrogenation using sulfur and halogens. *Sekiyu Gakkaishi*, **16**(5), 426-9 (1973).
15. F. Wessely, F. Grill, Dehydrogenation of organic substances with sulfur. *Monatshefte fuer Chemie*, 77, 282-92 (1947).
16.   US   DOE   2011   Interim   Update   for   Hydrogen   Storage. http://www1.eere.energy.gov/hydrogenandfuelcells/mypp/pdfs/storage.pdf

Author Index